A THEORY OF EVERYTHING

Published under licence by Brown Dog Books and
The Self-Publishing Partnership Ltd, 10b Greenway Farm, Bath Rd, Wick,
nr. Bath BS30 5RL

www.selfpublishingpartnership.co.uk

ISBN printed book: 978-1-83952-459-2
ISBN e-book: 978-1-83952-460-8

Cover design by Kevin Rylands
Internal design by Andrew Easton

Printed and bound in the UK
This book is printed on FSC certified paper

A SELF-DEVELOPMENT BOOK FOR EVERYONE

A THEORY OF EVERYTHING

Things I Wish I'd Known at Your Age

DR JONATHAN R. TUCK

BROWN
DOG
BOOKS

Table of Contents

A Theory of Everything

Dedication

I dedicate this book to the late Ruth Davies, my surrogate mother, who provided me with the understanding and mentorship that I needed as I was growing up, who gave me the platform that I needed to believe in myself, and who set me on the path to find my true success.

And of course, it goes without saying that I also dedicate this book to my beloved sons, William and Alexander Tuck.

Foreword

Dennis J Pitocco:

Having presided over an award-winning global new media digest over the past decade plus built upon thousands of literary works from around the world, I've had a birds-eye view of essays across the life, culture and business spectrum.

Rarely have I ever come upon such an extraordinary collection of insight and inspiration such as that discovered with this book. Jon has clearly and convincingly distinguished his work from so many others via a mastery of the written word, coupled with a refined knack for keen observation and a refined art of storytelling.

What you are about to read isn't just another book, but rather a living, breathing guide which you can reflect upon and embrace as a compass for changing your life's direction. It forced me to step back, ponder, and pivot in a more thoughtful direction. Brace yourself, as you will no doubt do the same.

Dennis J Pitocco
Founder, Publisher & Editor-in-Chief
BIZCATALYST360.COM

Chief Reimaginator
360° Nation

A Theory of Everything

Chief Encouragement Officer
GoodWorks 360° Foundation

Contributing Author
Chaos to Clarity (Amazon Best Seller

Céline Cloutier:
A Theory of Everything is an amazing book by an amazing man.

It is a book as vast as the ocean and limitlessly filled with knowledge, life experiences, references, answers, insights, inquiries and actually offers samples of all that one is required to reflect upon, to awaken to what truly is: YOU.

Jon's endless curiosity for life, his deep experiencing of what suffering is, self-realization and research, place him in an honourable position to share his truths.

Jon's guidance comes from genuine and unconditional love for humanity.

This book holds infinite life knowledge, not learned from education, family, religion, science, or traditional medicine.

This knowledge comes from the unlimited natural intelligence of rhythm of nature, available to all, yet for most to be awakened.

Quote from Jon:
"So, what makes me think that I have the knowledge or authority to describe to you, what I think life is all about and how to get the best from it? Well, I guess after reading this book, you will have to agree that I've lived quite a bit for a start! And my life hasn't been the smoothest of journeys. In fact, it's been so rough at times that I am surprised I am still alive to tell the tale."

Couple of Jon 's NEAT LIFE HACKS:

"Just thinking about what others might be saying or thinking about you in a negative respect is negative self-thinking, and if you're not careful, in this way can become a self-fulfilling prophecy."

"If you're not putting yourself out there and making mistakes, and learning from them, then inevitably, you're not moving forward in life."

I highly recommend this one-in-a-million book to parents, students, grandparents, especially to individuals who are looking for answers to their daily life challenges.

Céline Cloutier:
Co-author of TREES HEALING and YOU & Guide for Vibrational Tree Essences
Co-founder, Ambassador, Consultant of Tree Essences, Canada
York University, Toronto, Canada Graduate: Dance Therapy
Université de Ottawa Graduate: B A Psychology
University of St Paul, USA: Certificate of Conflict Resolution and Mediation
Reiki Association: Master's Degree

Nadeem Lutfullah:

It has been a while since I came across Dr Jonathan for the first time through a mutual friend, and to be honest, **I was immediately struck by the fact that he had a unique perspective on the world, and human life in particular**. I find such people extremely interesting and inspiring.

During the course of time, I have found Dr Jonathan to be a man full of passion, **with a profound love for humanity, its wellbeing, and its future**. The title of his book couldn't have been more appropriate than *A Theory* *of Everything*. Once you start reading it, you notice the straightforward honesty and the very candid narration that characterizes his style.

He is someone who overcame incredibly difficult and challenging times right from his early years. You might think that there are millions of other people who have gone through hard times. But what puts Dr Jonathan in a prestigious group of people is his quest to educate readers about what 'life' really is, and why they should be aware of facts that most of us spend our whole lives oblivious to.

In my humble opinion, this book will continue to serve and guide humans for a long time to come, benefiting people of all age groups, cultures, and backgrounds. **It will be my privilege to learn more about Dr Jonathan's amazing research about the world and its habitants. This amazing book also encourages introspection, a willingness to examine one's own feelings, thoughts, and actions**.

I end with one of my favourite quotes from his book, which illustrates the importance of 'failures' in our lives. A concept that we do not often get to hear or read about.

"If you're not putting yourself out there and making mistakes, and learning from them, then inevitably, you're not moving forward in life."

Nadeem Lutfullah
Career Mentor & Coach | Teacher/Trainer | Aviation Veteran | Human Servant | Lifetime Student.

Jill Sudbury:

Jon is a wonderful balance of research-based scientific cleverness; his brain is like the planet. He has the rare combination of creativity, and love and understanding. He's able to balance both hemispheres of that brain of his and, for example, create a rainbow bridge for us all to enjoy and understand! Our learning, meaning and understanding is much enhanced by everything he shares.

He's lived with great pain due to his childhood traumatic accident, which I believe has given him the rare gift of insight, motivation to heal himself, and to share it with those around him. He's overcome addiction, and loss, and survived it with great strength and wisdom.

This book is a map of life, meaning and everything, to share with Alex and William. An absolute joy to read as a guide filled with love and truth, with quotes, references, and a universe of love and wisdom.

This life guide, full of truths and humour and raw honesty, suggests that the boys would be well to find research before jumping to conclusions, and teaches them to learn and to enthuse.

It will appeal to all parents, young people, singles alike and I'm sure that it will provide answers and indeed questions to be asked for generations to come.

Against all odds, it is said that we are entering a golden age of knowledge, wisdom, and above all, love. Love is all, and this book is full of it in its pure and glittering essence. I love this beautiful soul, even though his autobiography's titled: *Who the F*ck... is Jon Tuck*...Enjoy!

Acknowledgements

I do not know if what I'm going to tell you will be *your* truth, or whether it will become your truth, or if it even *should* become your truth, but I humbly acknowledge that the value of writing this book to you has been of tremendous benefit to me as a person. I want to thank you, and everyone who has made this possible, and who has helped me along the way.

I'd like to thank:

My twin flame Angela Mary Moore, and with gratitude to the Universe for preparing us for seven years since our first connection, where we found ourselves to be in a position to accept each other harmoniously, peacefully and through perfect synchronicity. I'm certain that the Universe brought us together for a specific purpose, to fulfil our unified quest to play our part towards the path for humanity, and ultimate ascension at this time. Needless to say, we manifested each other consciously, and at the right time for our relationship to be ultimately successful, and with the potential to achieve the most magical accomplishments in our current lifetimes on this earth. Many thanks and gratitude to Angela for her support and assistance with editing this second edition of my book, and in a way that it is more acceptable for international audiences.

Dr Linda Mallory, for providing me with the inspiration to write this book, and for the invaluable advice and unbiased wisdom that she has shared.

Roger Davidson, for introducing me to energy-based healing equipment, such as Rife technology, and other equipment originally designed by the great Nikola Tesla, and for providing me with his uniquely researched perspective on elements of science, which are very much overlooked in the scientific community and society today.

Chris Davies for his invaluable assistance in the early stages of setting up and using my pulsed electromagnetic frequency healing equipment, for the exciting journey that he's shared with me developing vortex-based healing technology, and for otherwise expanding my universe.

Clive de Carle for assisting and educating me on the truth surrounding many elements of health and wellbeing and the most succinct routes to recovery, and for introducing me to diathermy equipment, and for helping me with Nikola Tesla violet ray equipment.

Alistair Savage and Jackie Tait (née Skinner) for being my best friends as I was growing up in my teens, and to this day. Al, amongst other things, for providing me with 'the Al Savage guide to street credibility', and Jackie for giving me the confidence that if I wasn't married by the age of 30, then she'd do it for me!

Ian Whiting for being such a good friend both personally and in business, and for providing me with the feedback on the first draft of this book, and for spitting his coffee on the person sitting in front of him on the train when he was reading it (laughing)!

Marcel Alexander Buth, for being such an amazing and inspirational buddy in China, and for providing invaluable support to my editor and me in the early days of construction of this book.

Justina Miko for her help in constructing and conveying my understanding of 'the significance of our thoughts', and Jarek Mikolajczyk for support and advice, and some very interesting discussions.

Richard Davies, for giving me the most profound piece of advice early on in my life.

Robbie Holder, for being such an accommodating and generous friend when I was a student. Please know that I will never forget that I have promised to repay that debt of gratitude to you.

Jill Sudbury, for synchronously appearing and supporting me at a time of need, and for engaging me in her beautiful community. For subsequently training me to be a Reiki teacher, and for inspiring me and entrusting me to develop a workshop explaining energy healing from a quantum-energy perspective. Jill, you are a much-loved inspiration to all in your community!

Michael Heritage, for being a truly divine being and ally, for everything he's so selflessly done for me, and for joining me and assisting me in my pulsed electromagnetic frequency healing journey.

Professor Alexander M Korsunsky, for providing me with the most profound mentorship and guidance throughout my Doctorate, and for taking me to Trinity College and the University of Oxford with him.

Chris New, for introducing me to what I now understand surrounding energy and the overwhelming and very under-utilised power of the mind in our 'modern' society.

Herman Susanto, for being such a great mentor to me, when I needed it the most, during my time as a Business Coach.

Jill Sudbury (Goddess), Michael Heritage (Archangel), Lynne Mawby, Hillary Swift, Chris Fuller, Alex Loveday, Hilary Swift, Julie Carnelli, Lou Cook, Shanni, Cecelia Phillips, Lynda Gillingham, Jamie Riggs, Kimberly Groves and Ian, Jessica Sudbury, Amethyst Grey, Ivan Brownrigg, Kay van Beersum, Andy Potter, Ann Griffin, and Adam Boonham (and many more) at the Marlborough School of Reiki, for the invaluable support that they've provided me with, the beautiful community that they built (primarily by Jill Sudbury), and for the wisdom that they taught me when I needed it the most.

Julia Edwards for the abundance of enlightening and interesting discussions, and for the plethora of links and references that have substantiated my research and knowledge on a range of topics over the years.

A Theory of Everything

Hannah Gage (née Tuck), for supporting me and guiding me in the early stages of my energy healing journey, and for being there for me when I needed her the most.

Stephen O'Donnell, who amongst many other things, gave me my first insights into steps towards being a great father.

Julian Felstead, who took me under his wing and taught me about the world of business, when I was at a very low point, and enabled me to grow tremendously as a result of his support.

Paula Reed for introducing me to EFT and for helping me during a very challenging period of my life.

Sonia Ricotti, for providing me with the resources to change my life at exactly the right moment in time, and for making me truly 'Unsinkable'.

Christine at Lighthouse Readings for her clairvoyant advice and guidance over the past several years.

Stephen Voller, Tim Walder and Marappa Rajendran, for inviting me, and joining me on an exceptionally enlightening and fruitful journey, in starting a credible and valuable technology business. Stephen, I learned a tremendous amount from you, and I am extremely grateful.

Alicia Sawaya for being my spiritual herbalist, for the knowledge and information she's shared, and for the exceptional references and resources she's provided.

Brian Caswell, for being a part of my first Think and Grow Rich workgroup, and for providing me with 'Clarity', when I needed it.

Maciej Kuchcinski 'the Chemist', for educating me in the truth behind the chemistries surrounding so-called narcotics and hallucinogenics, and for being such a great buddy in China.

Bogumila Dorota Wolanska for introducing me to myself, for providing me with an insane adventure that I will never forget, and for helping me to finally understand and defeat the addiction we call alcoholism.

A Theory of Everything

To all the students and friends who have invited me and trusted me to mentor them over the years.

In no particular order: Sara Yadollahvand, Alistair Johnson, Roger Seawright, Mo Spires, Oliver and Zoe Spires, Robin Young, Ankara, Les Evans, Diana Davidson, Wayne Barrowman, Ralph Schussler, Xu Rai, Oscar and Fiona Hopkinson, Martin Davies, Emma Mullis, Ali Jones (née Tuck), Omar Bahadur, Janine Greaves, Paul Franklin, Dave Ford, Dave Barnwell, Pippa Thyne, William Young, Pritesh Hiralal, Restituto Castor, Amora and Damayanti, Henning Uhlenhut, John Shields, Paul Blackmore, Bruno Pollet, Jason Smith, Robert Trezona, David Rowe, Steve Farmer, Andy Mallon, Gleb Ivanov, Alex Inrig, Nick Lever, Anish Gami, Valerie Self, Suzanne Doherty, Jim Johnson, Emma Taylor, Mr Ashby, Mark Cary, Adam Dobson, Jason Smith, Matt Long,, Moon, Liuning, Iskra & Kristomere, Stoyko Stoykov, Rob McNaught, Lynn Peacock (blood brother), Mike & Valentina Parsley, Nick Johnson, Paul Artes, Prof Trevor Page, Prof Steve Bull, Prof Geoff Gibson, Prof Howard Chandler, Prof Pete Dobson, Prof Richardson, Richard Cooper, Saundra and Keith Artes, Lennie and Margaret Peacock, Sarah Wolkowski, Laure-Anne Visel, Scott Coleman, Sean English, Shufang Ma, Steph Shucksmith, Donna Grooby, Suzanne Gillhard, Catherine Carr, Quentin Lemarie, Richard Munro, Nazanin Rashidi, Andy and Penny Tuck, Anthony Mullins, Barry Grant, Mike Rigg, Ben Sheils, Chim Chu, Tarja Koivekko, Wayne Thomas, Colin Arnold, Con, William Owens, Zoe Spires, Ken Forsdyke, Daniel Ghadimi, James Ison, Stephan King, Dorota and Mariusz, Florence Leroy & Carl, Graham and Sheena Cherrington, Ahmet Erlat (Gun), Helen and Jim Rose, Mariana and Ilian, James Hammond, Jamie Tweedy, Jo and Jason Kent, Jonathan and Sarah Wood, Jules Christmas, Kevin and Sharon Lightfoot, Lisa Kennah, Louise Triance, Maggie, Emmanuel Eweka, Martin Hogarth, Ben Sheils, Chris Kidd, Mike Barlow, Bill Beaulieu, Ian Merrell, Karolina and Maria Swierczynska, Adam Sayed, and many, many more…

A Theory of Everything

In no particular order (except the first few): Nikola Tesla, Joe Dispenza, Bruce Lipton, David Icke, Greg Braden, Lynn McTaggart, Marisa Peer, Caroline Leaf, Stephen Hawking, John Lennon, Eckhart Tolle, Bob Proctor, Michael Tellinger, Dolores Cannon, Kurt Cobain, Jim Rohn, Sunil Bali, Noam Chomsky, Carl Jung, Keanu Reeves, Jim Carrey, Napoleon Hill, George Harrison, Robin Williams, Bill and Esther Hicks, William Wallace (in essence), Anthony Robbins, John Assaraf, Keith Flint, Oliver Reed, Brian Tracy, Ben Elton, Ernest Hemingway, Irvine Welsh, Phil Harris, Sig Hansen, Russel Brand, Lewis Hamilton, Joe Vitale, Deepak Chopra, Burt Goldman, Bob Marley, Wayne Dyer, Chris Cornell, Michael E Gerber, Bill Harris, Jack Cranfield, Richard Branson, Quentin Tarantino, James Hunt, Aldous Huxley, Michael Caine, Brad Sugars, Jim Kwik, David Wilcock, Ozzy Osborne, Leonardo Da Vinci, Phil Anselmo, Michael Losier, Napoleon Hill, Steve Jobs, Julian Assange, Wilbur Smith, David Bowie, Osho, and many more, who've provided me with much food for thought and invaluable information and insights throughout my journey, and continue to do so.

To my mother and father, and stepfather, for being my parents and for supporting me at every challenging period in my life, and for putting up with me being a very naughty boy (and beyond). To my mother for being such an inspiration to me, and for providing me with the invaluable mentality of 'glass half full' at such an early age, and for her unjudgmental ear and relentless support that she provided me with on my journey. To my father for showing me unconditional kindness.

And, even though we're not together, I especially want to thank your mother, for everything she did for me, and for everything she put up with from me, for all the information that she's provided me with on parenting and 'ADHD', and for investigating and involving me in the parenting workshop that we attended, which was given by Dr Linda Mallory, which inspired the construction of this book.

Disclaimer

In the presentation of this text to you, all of my ideologies and beliefs have been derived from a lifetime of observation, research, reading and direct experience.

I'm not perfect, and I'm not classically educated in human psychology, and I wouldn't consider myself a quantum physicist either. I don't have all the answers, and where I've presented anything as being 'the truth', it's what I personally understand it to be, to the best of my knowledge, as a result of the accumulation and compilation of my research, reading, self-education, experience and intuition. I fully accept that it is my truth, and that elements of it may not be right for you, or for anyone else.

"I once heard a wise man say there are no perfect men. Only perfect intentions." Morgan Freeman

---

Some of my views are in direct conflict with what many people have been taught and have adopted to be their truths. It's not in my interest to question their own research or level of knowledge, or truths. However, if they can't substantiate their beliefs, other than what they've been told to believe; then I don't care what they think, and I would encourage that neither should you!?

These are my truths. Truths that resonate with me, and truths that make sense to me. I'm happy to stand by them until proven otherwise. I wholly accept that some of what I'm presenting is pure theory, and that it's not quite perfected yet! We are forever evolving beings, after all.

However, I'm utterly confident that I'm not alone in my beliefs, and I know that there are a great many people, and more and more every day, who resonate with me, and understand and are open to everything that I discuss with them on these topics.

Other than providing you with the best information, guidance, inspiration and food for thought that I can, what I also want to achieve with this book is to help unite humanity on the principle of the greatest energy force in the Universe, that of **unconditional love**, and to help to change the world for a better and more harmonious place. A place that we can be collectively proud of to live in.

"Never doubt that a small group of thoughtful, committed citizens can change the world. Indeed, it is the only thing that ever has" Margaret Mead

"Our alienation goes to the roots…We are born into a world where alienation awaits us. We are potentially men, but are in an alienated state…the ordinary person is a shrivelled, desiccated fragment of what a person can be. As adults, we have forgotten most of our childhood, not only its contents but its flavour; as men of the world, we hardly know of the existence of the inner world…The condition of alienation, of being asleep, of being unconscious, of being out of one's mind, is the condition of the normal man…between us and It [our true selves or soul] there is a veil which is more like fifty feet of solid concrete…The outer divorced from any illumination from the inner is in a state of darkness. We are in an age of darkness…We are all murderers and prostitutes…We are bemused and crazed creatures, strangers to our true selves, to one another' (see par. 123 of FREEDOM)."
R. D. Laing

A Theory of Everything

They are my friends; they are my allies; and together in unity, we will change the world!

"Here's to the crazy ones, the misfits, the rebels, the troublemakers, the round pegs in the square holes... the ones who see things differently – they're not fond of rules, and they have no respect for the status quo.... You can quote them, disagree with them, glorify or vilify them, but the only thing you can't do is ignore them because they change things... They push the human race forward, and while some may see them as the crazy ones, we see genius, because the people who are crazy enough to think that they can change the world, are the ones who do." Steve Jobs

"The beginning is near. When the earth is ravaged and the animals are dying, a new tribe of people shall come onto the earth from many colours, classes, and creeds, and by their actions shall make the earth green again. They will be known as the warriors of the rainbow" Native American

---

References and Quotes

I've attempted to make reference to all of the key learning resources that I've utilised over the years, and I've also referenced many of the most profound quotes that I've come across, from some of the most 'successful' people to have walked on the earth in this era.

NEAT LIFE HACKS: CONSIDER THAT THE ONLY DIFFERENCE BETWEEN 'SUCCESSFUL' PEOPLE AND OTHER PEOPLE IS THEIR **MENTALITY**, THE WAY THEY THINK, AND THE CONFIDENCE, DRIVE AND DETERMINATION THAT THEY HAVE IN THEMSELVES. I BELIEVE THAT ANYBODY CAN BE AS SUCCESSFUL AS THEM (ON THEIR CHOSEN PATH), IF THEY REALLY WANT TO, AND IF THEY SOUGHT THE TOOLS TO DEVELOP THEMSELVES TO BE THE WAY THEY NEED TO BE TO SUCCEED

"Success is when we wake up and notice that the true purpose of life is simply being yourself" LINDA MALLORY

NEAT LIFE HACKS: JUST BECAUSE SOCIETY TEACHES US TO VIEW QUOTES AND SAYINGS AS 'OLD WIVES' TALES' OR 'CLICHÉ', AND ENCOURAGES US TO PAY LITTLE ATTENTION TO THEM, IT DOESN'T MAKE THEM ANY LESS PROFOUNDLY TRUE OR BENEFICIAL TO THOSE WHO ARE BRAVE ENOUGH TO INDULGE THEM

I'm conscious of the fact that there's some controversy surrounding whether these famous people actually made the quotes that are being attributed to them, and whether it is, for example, just their 'marketing people' using the quotes for publicity purposes. Frankly, I don't care either way. I've referenced the quotes because they make sense to me and they resonate with me, and they concisely explain what I'm trying to convey to you!

I realise that the live information links that I've added to the text may not remain live links forever and may change over time. This is the unfortunate side of our ever-evolving world wide web. I will endeavour to update the links given in this book regularly. However, hopefully the text surrounding the link should provide sufficient information and inspiration to either help you find your own further information, or will give you the insights you need.

One very good resource I've used previously, which you may find useful if a link disappears, is called *The Way Back Machine (https://web.archive. org/)*. This is an archive of every website ever published, so you should be able to find anything here that I've referenced.

Prologue:

To my beloved children, William and Alex.

This is the beginning of a book that my whole existence on this planet earth, or at least my life to date, appears to have been leading to; and I write it for you.

One of the many incentives that I have in writing this book are some memories that I have surrounding my behaviour towards you when you were very young. I've learned to understand how your memories of my behaviour may have scarred your subconscious, and moulded, for example, your anxieties, confidence and demeanour. I have no excuses. Just my own reasons. Reasons that I eventually used as a foundation to address my psychology and behaviour.

It's become my mission to learn how I can put myself in a position to be able to heal those scars in both of you in the future. This has been the main drive behind writing this text.

You see, one of the things I always prided myself for, is that even when I had been drinking, I very rarely resorted to physical violence. But what I subsequently learned about verbal violence shocked me to the core, and deeply disturbed me, and dictated the challenges that I tasked myself with, to make good for the future. Many thanks to Emma Mullis for this initial insight. Blaming myself or beating myself up about the past isn't going to help anyone, least of all myself, but I can hold myself accountable and responsible, and I can research and investigate, and practise everything I learn, to attempt to rectify the damage that I may have inadvertently done to you.

One of the most profound memories that I have of my displaying such vulgar behaviour was when I was going through the process of starting my first official business, and then my first (and I expect, last) bankruptcy. It was a dark and challenging time for me, from which I was to gain a plethora of invaluable lessons and wisdoms.

I remember that Alex, you were about two years old and had just started walking. You were on the stairs, and I remember for whatever reason being very angry with you and shouting at you until the spittle projected from my mouth. I was pulled away and told to calm down, which I did, before rationalising with myself, and with you, and calming the situation. It was over in a flash, and calmed as if nothing had happened.

I think what made the particular event more poignant, though for all of the ten seconds that it lasted, was the fact that we had some very close friends to stay, who subsequently commented on my bad behaviour. This cut deep and hit home significantly, and it made me realise that I'd behaved very badly, and that I couldn't use any excuses, and that I had to do something about it.

And the more I later learned about childhood trauma and emotion/ memory/ sub-conscious, the more I understood just how significant that event, and others were, for you both, and your future development and outlook on life. What I was to understand was how the memory of that event may have deeply scarred your subconscious, and moulded, for example, your confidence and demeanour. It gradually became my mission to learn how I could put myself in a position to be able to heal those scars in both of you in the future.

I'm grateful for the mistakes that I made, and for the support that I received, and for the opportunity this presented me with to do the significant research that I've done, and the work that I've done to address this goal, and many others, which I can now present to you, my boys.

Firstly, let me tell you this (and I say this without too much bias, being your father):

You are both special. *Very* special. You are both unique, and you both have the ability to be very happy, fulfilled and successful in life, in whatever it is you want to do, see, experience, or become.

You are both exceptionally intelligent young men. My intention is to guide you in the significance in being able to **think** for yourselves, and to help you to begin to develop your **wisdom**. But I can only *help* you to begin, and to help to provide you with the best guidance I'm capable of throughout your journey, if you want it. The rest you'll have to figure out for yourselves. By learning from your mistakes, from the questions that you ask, and the answers that you trust. From whom you engage with, from the **intuition** that you use, from how you filter the information you get from the research you do, and through the experience of your own lives as you grow older.

"To be old and wise, you must first be young and stupid"
SunGazing

I want to help you to see that you have a **choice**, right here and now, and throughout your life, as to whether you see your journey as a continual test and an arduous uphill battle, *or* whether you see it as a wonderful, inspiring and fulfilling adventure. I know it can be the latter, and I know that you can experience it that way, if you so choose.

"The 3 Cs of life: Choices, Chances and Changes. You must make a choice to take a chance, or your life will not change" *Zig Ziglar*

A Theory of Everything

"I Choose... To live by choice, not by chance. To be motivated, not manipulated. To be useful, not used. To make changes, not excuses. To excel, not compete. I choose self-esteem, not self-pity. I chose to listen to my inner voice, not to the random opinions of others" Unknown

"The Secret to Life... is that you have all the power. If you can change your mind, you can change your life. You don't need any amount of money, education, or connections. You just need to be in control of your mind. Program your mind to believe in yourself. Know that you have what it takes to achieve greatness, and watch greatness manifest in your life. When it comes to the law of attraction, all you're really doing is programming your mind. This is the secret that so many people overlook in life. The key to success is overcoming your limiting beliefs and aligning yourself with the energy you want in your life"
Third Eye Thoughts

You're also both gifted with good looks, and although I'm of the belief that, as human beings, we should all be seen and treated as equals, in the society in which we live, ultimately your looks may give you some good advantages.

The media, society in general, and people all throughout your lives will tell you what they think you should be thinking, or what they think you should be doing, and how they think you should be doing it. They might be extremely persuasive and insistent, but I encourage you to have faith in yourselves, to follow your own intuition and hearts, and to do whatever it is that makes you happy in the long run, as long as it's done with kindness to others.

A Theory of Everything

"If you end up with a boring miserable life because you listened to your mom, your dad, your teacher, your priest, or some guy on television telling you how to do your shit, then you deserve it"
Frank Zappa

"Your time is limited, so don't waste time living someone else's life" Steve Jobs

"Wanting to be someone else is a waste of who you are"
Kurt Cobain

"Can you remember who you were, before the world told you who you should be?" Danielle LaPorte

---

Nobody, including myself, could possibly advise you ultimately on how to live your lives any better than you can guide yourselves, and to know your own truth. Follow your intuition and your dreams, and all will become revealed in time.

NEAT LIFE HACKS: THEY SAY THAT A DREAMER IS A FOOL. I SAY THAT IF YOU DON'T FOLLOW YOUR DREAMS, THEN YOU ARE BY FAR THE GREATER FOOL

"If you can dream it, you can do it" Walt Disney

"Dreams at first seem impossible, then seem improbable, and finally, when we commit ourselves, become inevitable"
Christopher Reevei

"Anything is possible" James Hunt

A Theory of Everything

NEAT LIFE HACKS: IF IT'S WHAT YOU ENJOY DOING, THEN IN GENERAL AND WITHIN REASON, IT'S PROBABLY WHAT YOU **SHOULD** BE DOING, AND WILL INEVITABLY LEAD TO YOUR SUCCESS AND HAPPINESS, AND THE HAPPINESS OF OTHERS AROUND YOU, IN THE LONG TERM

"The reality is that no one is actually completely average and has at least one skill or talent, be that as yet undeveloped, which is well above average. In a world that has perpetuated the cult of the average and valued sameness, conformity is losing its grip on the reins as authenticity and exceptions rule. There's one thing that you're a world champion at. No one does it better than you. You're the best at being you. And when you're being your best self, your world will transform from a round hole to the shape of your square peg" Sunil Bali

---

I'm telling you this from what I would consider to be good experience. I spent way too much time in my early life and career listening to what others told me they thought I should be, do, or have, instead of filtering out all the nonsense and trusting my own intuition.

Some people come into your life with good purpose and will teach you everything you need to know at that point in time. The key is in being able to recognise who these people are, who you can trust, and to open yourself up to learning from the experience.

The best advice I can offer you in this respect is to **trust your intuition, or your 'gut feeling'**.

We are born with this very powerful guidance tool which is often overlooked, and which science is now even looking at on a biochemical level (see chapter 9). But if you follow your intuition or 'gut feeling' you're probably on the right track, even though the outcome may not be what you were expecting at the time.

A Theory of Everything

"Instinct is something that transcends knowledge" Nikola Tesla

NEAT LIFE HACKS: OFTEN WHEN FACING CHALLENGING SITUATIONS, THE BEST PATH THAT YOU CAN FOLLOW IS GUIDED BY YOUR 'INTUITION' AND 'GUT FEELING'. I BELIEVE THAT THIS IS ONE OF THE WAYS THAT YOUR SOUL, AND INDEED THE UNIVERSE ITSELF, IS ABLE TO COMMUNICATE WITH YOU, AND GUIDE YOU ON YOUR JOURNEY.

Part 1:
What Qualifies Me to Think I Can Explain Anything About Life to You

1.1 The Breaking of My Head

So, what makes me think that I have the knowledge or authority to describe to you what I think life is all about, and how to get the best from it? Well, I guess that after reading this book, you'd have to agree that I've lived quite a bit, for a start! And my life hasn't been the smoothest of journeys.

In fact, it's been so rough at times that I am surprised I'm still alive to tell the tale.

This first part of my tale starts in the summer of 1982, at the age of eight years old. My friend Nikki and I were out on one of our usual bike rides. It was a Saturday and a beautiful day, and my family were due to go down to the coast as we often did during the summer months.

I'd been in trouble the week before because I'd got a puncture in my bike tyre, and my punishment was that I wasn't allowed to ride my bike, but that was alright because my nana (your great-grandmother) was in charge for the morning, and she was unaware that I wasn't allowed to ride my father's bike! My poor nana. I don't think that she ever really forgave me for that.

A Theory of Everything

So that morning I took my father's bike, and Nikki and I were doing our usual rounds of Steyning. We were coming down the steep hill with the park on the left at the bottom. As we neared the bottom of the hill, I remember that we started to back-pedal; we did this a lot when we were riding, and I'm not sure why we did it, but on this beautiful summer morning this habit was to change my life forever.

You see, I'd forgotten that my father's bike had a back-pedal brake and I skidded awkwardly at high speed. Back then, wearing a cycling helmet wasn't common, so when my head hit the kerb by the side of the road it connected with a crack and I knew that I'd hurt myself pretty badly. This was nothing new, though, as I was always hurting myself, with cuts and bruises a regular feature on my body.

I'd cut my left shoulder quite badly and the main impact was to the left side of my head, but strangely the pain was on the right-hand side. My mother bathed me when I got home, and later on she remembered where I'd said my pain was, and that single fact probably saved my life that day.

The rest is a bit of a blur and I can only construct what happened in my mind from bits and pieces, and from what people have told me about it afterwards.

The family decided to proceed to the beach as planned, but my condition worsened, and I started vomiting. We quickly made our way to Shoreham Hospital. Here I passed out.

I'll never forget my mother saying afterwards that she wished that I'd been awake for the journey down to Southampton neurological ward, as we were escorted down the M3 in an ambulance by no less than five police cars! The surgeon listened to what my mother had to say about the pain and immediately took me to emergency surgery. He straight away made an incision and drilled into my skull where I had indicated the pain had been. He uncovered a massive blood clot in my brain on the right-hand side.

A Theory of Everything

You see, what had happened was that when the left-hand side of my head had hit the kerb, my brain had bounced across and impacted against my skull on the right-hand side. This in turn had caused a massive haemorrhage.

After the operation I fell into a coma where I resided for two weeks. It must have been the longest fortnight of my parents' lives. They both took it in turns to stay with me and to read to me by my bedside. It was during a reading of 'Puck of Pook's Hill' that my father saw me stir for the first time.

The first thing I remember when I 'awoke' was the feeling of absolute and utter hopelessness. I couldn't move, I couldn't see, I couldn't talk, I couldn't communicate at all; all I could do was hear and think. I was scared. I was conscious, but I was trapped inside a dormant body. In years to come when I listened to the Metallica song 'One' it would bring back memories. It was about a survivor of the Vietnam War who was very badly dismembered by a landmine and was left with nothing but a conscious mind entombed in a torso, just wishing for death. Strangely enough, this became one of my favourite songs.

All I could do was think, and all I could think of was the fact that I was desperate for the girl in the opposite cot to shut the fuck up! She was screaming and screaming and jumping around in her cot, and it was driving me nuts!

As the days went by, I learned to communicate by pointing. The right side of my body was completely paralysed, almost like a stroke survivor, so I had to learn to do everything with my left hand. Holding a cup and fork, for example, but I also had to learn to write with my left hand, something I can still do to a degree today.

Interestingly, it was explained to my parents at the time that, having suffered a head injury, it was likely that the first word that I would utter when I regained consciousness would be a 4-letter swear word. So, they all gathered around my bed eagerly anticipating the profane torrent that I was about to deliver. However, unlike you, Alex, I didn't know any swear words at that age, which left them all very disappointed. The first word I spoke was 'hello' to my sister, your Aunty Ali.

I was bald; all of my hair had been shaved off and I now had a scar on the right side of my head which to all intents and purposes looked like a massive caricature of a penis that we used to draw for fun when we were kids. I still push the tip of my finger into the spot of maximum indentation to this day and wonder what I am going to do if my hair thins much more or if I go bald. At least I can say that it's '*Dr* Dick Head' to anyone who might joke about it!

After the hospital days, I remember being stationed on a low mattress in my parents' room and having a bell to ring if I needed anything. I was given a 30% chance of surviving with a normal life ahead of me, the surgeons advised my parents not to expect that I would be anything more than a 'cabbage', which was acceptable terminology back then. I was going to prove them wrong!

1.2 Life's Best Gifts Sometimes Come Badly Wrapped

I think that because of the severe accident and injury to my head, I actually experienced a very good '*head*' start on many people in life!

Firstly, it gave me indefatigable determination, and ultimately the knowledge that if you really want something, then you can achieve it, despite the odds, and despite what anyone tells you, so long as you believe it and have faith in yourself. I notice this trait in you, too, William, as when you set your mind to something, you seem to dedicate your full attention to it and display great determination, passion and resilience.

Just look at what they told me back then in 1982. They said that I had a 70% chance of living the rest of my life as a '*cabbage*'. They also told me I'd be in a wheelchair by the age of 30. But in the years to follow I thought: '*Fuck them!* What do they know? They didn't know me, or who

A Theory of Everything

I was, or what I was capable of'!

And yes, I'd learned to swear by then, and enjoy contextual swearing (mainly) to this day!

Secondly, it gave me a platform very early on in life to develop my left brain-right brain balance. As a result of the paralysis I experienced in my dominant right side, I was fortunate to be forced to learn to use my left side to do everything, including writing with my left hand in the early days. To this day I've used my left side for many activities including holding my drinks, playing sports like darts, table tennis, playing pool, playing football and cricket, using the computer mouse, and smoking cigarettes.

On top of that, and for whatever reason, I've always been very positive, and able to see the good in any bad situation. This positivity has only increased as I began to realise everything I'm going to explain to you in this book.

I think it also made me very courageous. It enabled me to see gradually, that if an opportunity was there, and if it was reasonably safe, then I *had* to take it because it may never present itself again. And I was damned if I was going to *regret* not having taken it. My friends at university used to tell me that I had balls the size of planets. I think much of it may have been more accurately attributed to stupidity and naivety back then, but I accepted the accolade!

"Whether you believe that you can, or you believe that you can't, you're right" Henry Ford

---

In hindsight, I'm eternally grateful to the medical team for setting me those challenges back then. For setting the survival fire alight inside of me, and for ultimately giving me the strength to be who I am today. Oh, and I'm especially grateful that they saved my life, too. But that goes without saying!

1.3 What I Learned from my Number 1 Life Trauma – A New Outlook on Life

Another gift that I was blessed with was that I was acutely aware in my early to mid-teens that my outlook on life was likely to be hindered substantially by the psychological trauma I suffered at such a young age.

I could see that consciously, what I wanted to do, how I wanted to behave, and what I wanted to achieve, seemed to be blocked by something. I was determined to find out what and why. I needed to find a way of resetting myself, and to change my negative and destructive core beliefs, into positive and productive ones. Core beliefs that I was learning to understand were being held deep in my subconscious.

The way I eventually learned to see myself and the world stemmed from my early interest in quantum physics. I was introduced to quantum physics during my first degree and it bridged the gap between physical sciences (biology, chemistry or classical physics) and non-physical sciences or phenomena, such as religion, spirituality, the mystical, and the metaphysical.

"If you want to find the secrets of the Universe, think in terms of energy, frequency and vibration" Nikola Tesla

"The day science begins to study non-physical phenomenon, it will make more progress than it made in all the previous centuries of its existence" Nikola Tesla

---

Just because you can't see something, measure it, or quantify it, doesn't mean that it isn't a real, or an acceptable phenomenon.

My newly acquired knowledge about the energy around us eventually helped me to physically, metaphysically and even spiritually move forward and change my core belief system for the better.

If I hadn't had this trauma, or adversity, at such a young age, and throughout my life, then there's a chance that I would not have embarked upon this journey of discovery and enlightenment at all, and I would not have been in a position to understand and want to address the damage that I may have done to you as children.

I am sincerely very, very grateful.

1.4 Early Energy Beliefs

Let me explain: Everything in the known universe is simply **energy** at its fundamental level; atomic, photonic, quantum, light, sound, waves, and absolutely everything in existence, both physical and non-physical. And the way we perceive that energy through our senses depends upon its vibrational frequency, and the sensitivity of those senses.

"Everything is energy and that's all there is to it. Match the frequency of the reality you want and you cannot help but get that reality. It can be no other way. This is not philosophy. This is physics!" Albert Einstein

"Nothing rests. Everything moves. Everything vibrates. At the most fundamental level, the Universe and everything which comprises it is purely vibratory energy manifesting itself in different ways. The Universe has no 'solidity' as such. Matter is merely energy in a state of vibration" Unknown

"Everything we call 'real' is made of things that cannot be regarded as 'real'. Those who are not shocked when they first come across quantum physics cannot possibly have understood it" Niels Bohr

---

A Theory of Everything

So, the energy in a solid 'object' is vibrating at a lower frequency than the energy of a liquid or gas. Basic classical physics, right?

But this doesn't just apply to inanimate objects. It applies to absolutely everything in the universe, including the matter that constitutes plants, animals, human beings, thoughts, and space itself. This is something we aren't explicitly taught at school. Also, human beings, as wonderful and miraculous as our sight is, have a clearly limited visual spectrum. Equally, our hearing is limited. For example, a dog whistle cannot be heard by most humans.

We can't see non-physical energy, among other known phenomena, such as radio waves, atoms, magnetism, dark matter, antimatter, Wi-Fi, toroidal energy forces, the oxygen and most other gases we breathe, ultraviolet and infrared light, gravity, the mind, and God; we *can't see God*. And equally importantly, we cannot see sound, emotions, thoughts, or love. But we accept that they all exist. Or should I say for the sake of clarity, we accept what society teaches us is acceptable *for us to believe exists*. In fact, even thoughts and emotions are energy and 'real' and tangible things, and can be transmitted through space and time.

To complete the puzzle at this stage, what I learned was that everything in the universe, at a quantum level, is connected to everything else by the very nature of the energy it is derived from, as I said, **including our thoughts**. The Universe would appear to be one immense symbiotic energetic organism!

Documentary Reference: Everything and Nothing: The Amazing Science of Empty Space (tinyurl.com\y9737uau)
Documentary Reference: What the Bleep Do We Know...? (tinyurl.com/y7suu3xu)
Documentary Reference: 200,000 Year Old Levitation Technology - Michael Tellinger (youtu.be/OO_EoqnabQ8)

A Theory of Everything

This understanding is what I've based some of the philosophies in the coming chapters on and forms part of what has provided me with the mental support to heal and flourish as a human being. For those interested in enhancing their own journey and embracing the concepts of energy and quantum physics, a couple of exceptionally powerful and enlightening books that I read on the topic are *The Field* by Lynne McTaggart, and *The Biology of Belief* by Bruce Lipton. Also, the documentary, 'What the Bleep Do We Know...?' (referenced above) was of great value.

Book Reference: The Field – Lynne McTaggart (tinyurl.com/ybsta9sd)
Book Reference: The Biology of Belief – Bruce Lipton (tinyurl.com/y8lsj8sg)

One of the things that Grandma told me when I was a child was that *theoretically* teleportation is possible! This statement already had a profound impact upon my desire for knowledge and information on this topic.

If everything at its fundamental state is made of energy that is essentially light waves or particles, then it is theoretically possible that, by some means, we could envision the ability to take them all apart, transmit them across space and 'time', in the same way we send radio waves, or other communication waves, and then reassemble them at the other end. This is a well-documented phenomenon, and the subject of many interesting films, such as *The Fly* and *Star Trek*.

In fact, I've found that if you look at the subtext behind many of the great movies, a lot of what I'm gradually understanding to be 'the truth' is actually presented in plain view for people to see, such as through films like *The Matrix*, *RoboCop*, *The Shining*, *The Dark Knight*, *The Hunger Games*, *Lucy*, *Doctor Strange*, *Avatar*, *Inception*, and many of the other Marvel movies, and many Disney or family movies, including *The Croods* and *Moana*. These are just a few examples, but the list is endless. The more I learn about what I believe to be the truth and actualities of life and society, the more I recognise the subtle *truth* messages that seem to be conveyed in mainstream movies. I urge you to watch some of these movies and decide for yourself.

Please see a very good description of the messages being conveyed in mainstream movies, such as the ones mentioned above, provided by David Furlong in his book, *Illuminating the Shadow*.

Book Reference: "Illuminating the Shadow", by David Furlong (tinyurl.com/ybxblhw4)

Also, please see Collective Evolutions interview, "Disclosure and the Fall of the Cabal' with David Wilcock", which explains his understanding of how these messages are entering the mainstream, and why:

Documentary Reference: 'Disclosure and the Fall of the Cabal' with David Wilcock (tinyurl.com/y7pz9bfb)

1.5 Introduction to Energy Healing

Back in the early days of my investigation on how to reduce the permanent pain I'd been left with from my childhood accident, I tried everything I could find on the open market, including alcohol, medicines, both conventional and alternative, physiotherapy, acupuncture, osteopathy, chiropractic treatments, drugs, etc. I became quite desperate because nothing appeared to work or have long-lasting effects.

At that time, a friend of mine, Chris New, who was once known as 'The Shaman', and who I knew to have much experience and knowledge in energy healing, came to me and said:

'I've done it! I've worked out a method for immediate and permanent removal of pain from anyone, and I need another guinea pig to test my method on, because it doesn't work with my partner.'

Well, at a point of desperation, and having tried everything I knew of, I had nothing to lose, so I said, 'OK, let's crack on with it!'

To cut a long story short, what he was able to do for me was quite profound at that time, but what I found most remarkable was his ability to access my mind and remove memories instantly and tangibly, and then replace them with new beliefs, **over the phone, as I was talking to him**.

It literally blew me away.

He'd say: 'think about (whatever memory it was that was associated with what we were trying to address)', and then a few seconds later, he'd say '*now* think about it'. And bugger me, the memory was completely erased, and I could no longer think about it! In this way, if only temporarily, he cured my pain.

As an aside, this is also how he cured my fear of heights. I don't know if you remember, but your mother took us up The Shard, when it was first opened, to test whether what he had done had worked. It had worked so well that I was thoroughly bored of the whole experience after about half an hour up there!

He wasn't able to cure my fear of public speaking at that time, which was unfortunate. But I'm eternally grateful to him for addressing my pain, curing my fear of heights, and, more importantly, for introducing me to, and demonstrating to me the tangible effect of, the power of the mind, energy healing, and remote 'energy' healing and intervention. It had worked and all I could do was believe in it.

I was a rubbish guinea pig, though. All I wanted to do was understand the principles and mechanisms behind what he was doing, and he was reluctant to tell me, for whatever reason. The basis of what he was able to explain was that *he was just a facilitator, enabling people to allow themselves to heal themselves, and that the key principles involved were **intention, acceptance and allowance**.*

This knowledge, experience, and basic understanding formed the foundations of what I was later on to research in depth, and what inspired me to learn Reiki, a subtle yet effective form of energy healing using the hands to guide life force energy into the self and others.

Even though we rarely communicate these days, Chris, I am extremely and sincerely grateful to you for introducing me to the wonders of energy healing, and the Universal power of the mind.

1.6 Deep Breath!

OK, so I realise that I've probably just thrown some pretty heavy shit at you! So I'll try and explain it in small, bite-sized chunks as I tell my story. This should make it easier for you to understand, and should help you to realise that it's not scary at all (even quantum physics!). When you understand the power of your mind, and your thought vibrations, and when you take the effort to read or research for yourself some of the links I've put throughout this book, you can use all of this knowledge to fulfil yourself, find your happiness, and establish your ultimate success.

"Knowledge Protects; Ignorance Endangers"
"You can ignore reality, but you can't ignore the consequences of reality"
Aeon of Horus – The Occult History of NASA

NEAT LIFE HACKS: WHEN CHALLENGED BY A SITUATION OR TASK THAT'S WAY TOO BIG TO CONTEMPLATE ALL AT ONCE, THEN BREAK IT DOWN INTO SMALL STEPS, OR 'CHUNKS OF ELEPHANT', AS I WAS ONCE TAUGHT. WHEN YOU BREAK A TASK OR GOAL DOWN INTO 'BITE-SIZED' CHUNKS, AND TAKE THINGS STEP BY STEP, IT NOT ONLY HELPS YOU TO AVOID THE FEAR OF NOT BEING ABLE TO COMPLETE WHAT ON THE FACE OF IT SEEMS LIKE AN IMPOSSIBLE TASK, BUT ALSO

A Theory of Everything

MAKES THE WHOLE PROCESS IN GENERAL FAR MORE SIMPLE, AND FAR MORE
HARMONIOUSLY ADAPTABLE

So, going back to the purpose of me writing this text for you, I'm damned if I'm not going to try to save you guys as much time as I can in your figuring it all out! I mean, I think it's almost critical for your development to take the time to make the same mistakes as I did (except breaking your head – I wouldn't necessarily recommend that), but my intention is to help to speed up your learning process, so it won't take you the lifetime that it has taken me; *and is still taking me.*

NEAT LIFE HACKS: DON'T WORRY IF IT DOESN'T ALL MAKE SENSE TO YOU
RIGHT NOW, OR IF YOU FIND IT CONFUSING OR CHALLENGING, OR EVEN SCARY,
OR DON'T EVEN WANT TO KNOW ABOUT IT, OR THINK ABOUT IT AS A RESULT.
WHEN YOU'RE READY TO THINK ABOUT IT, IF YOU EVER ARE, YOU WILL DO
YOUR OWN RESEARCH, AND ONE DAY IT WILL ALL JUST CLICK

"For what it's worth, it's never too late, or in my case, too early to be whoever you want to be. There's no time limit, stop whenever you want. You can change or stay the same, there are no rules to this thing. We can make the best or the worst of it. I hope you make the best of it. And I hope you see things that startle you. I hope you feel things you never felt before. I hope you meet people with a different point of view. I hope you live a life you're proud of. If you find that you're not, I hope you have the courage to start all over again." F. Scott Fitzgerald

---

And let me tell you this now:

If you ever need, or simply want to know anything about anything, then do your research! Use the internet, and read books. Read lots of books! Don't think that your education is purely from what you learn at school. If you want to be properly informed, then you must self-educate, and the plethora of resources available to you provide the only challenge in as much as where you prioritise your learning, and what you believe to be true. My only advice here is to choose the information that resonates with you the most and compile as much information as possible before deciding your truth.

For when it boils down to it, as I see it, **your 'truth' is what you choose to believe.**

"The general population doesn't know what's happening, and it doesn't even know that it doesn't know" Noam Chomsky

---

1.7 Born with No Common User Guide

I'm utterly perplexed by the fact that it'd appear that we're born into this world, into the bodies, and with the DNA, genetics, and traits that we inherit, with absolutely no user guide whatsoever, and no given knowledge on how to live our lives, and what we're ideally meant to do, and whether we have a purpose, or not, or what it's all about.

I mean, when you buy any product on the open market, you expect it to come with a user guide or instruction manual. In many cases, we'd be completely lost without it. Even the most skilled of engineers is lost if this guide is absent, because he's got nothing to pretend to ignore!

When you think about it, the fact that we come completely without instructions sucks, doesn't it? Well, it did for me!

Like many people on the planet, I've spent what I can only assume to be half of my life (or, more accurately, my entire life to date) trying to figure it all out, and trying to reprogram myself to be at least a reasonable human being in my relationships with others, and in trying to be a good father to you, my two boys.

Wouldn't it have been a whole lot easier, and a better use of our time on this earth, if we'd have come into it with at least some insight into what we're here to do, and how to do it in the most harmonious, satisfying, succinct, effective and efficient way?

Well, actually, I believe that there are a number of elements to that statement that I've learned to understand, or at least accept, over the years. Mainly in terms of the significance of the mistakes we make, the lessons we learn, the hardship we endure, and the experience, strength, knowledge, confidence and wisdom that we gain from overcoming our challenges.

"What doesn't kill you, makes you stronger"
Neitzsche

Also, I'm aware of the 'user guides' that are available, having studied the texts from many of the religions around the world. Some Christians adhere to the Bible as their user guide. The Muslims use the Quran while the Jewish use the Talmud. The Tipitaka or Tripitaka is the guidebook of the Buddhists, while the Vedas is for the Hindus.

These are just a few of the guidebooks that some of us use to shape our truths, and I understand that they contain teachings about our purpose in life, which are generally not taught to us in schools, churches, or any other social groups that I know of. If you 'read between the lines' in these texts, you can find some very powerful truths and guidance indeed!

Wouldn't it be great if we were taught about finding our purpose by our parents, schools, churches, social circles and mass media, and about who we are, and about the powers of 'energy and intention' that we have available to us!

1.8 The Lost Memory of Ancient Civilisation

I'm convinced that if we, as human beings, had better memories, then whether for good or for bad, the world would be a very, very different place. It is utterly perplexing and debilitating to our species, that human beings, as evolved and as wise as we have the potential to be, with the most exceptional physical infrastructure, and the biological and chemical complexity that makes up who we are, have generally such poor memories!

I'd go even further to say that I think that poor memory is possibly one of the biggest issues that the human race experiences, in terms of its development as a species. I believe that our current condition of poor memory leaves us extremely vulnerable to our apparently gullible state, and appears to explain why history repeats itself again and again and again.

It's been interesting for me to watch documentaries, for example, which look at the chronological historical evidence behind something, to realise that it is so easy to manipulate and perverse the opinions and minds of the masses, because we simply can't remember the truth behind what's happened in the past. And it's not only the opinions that are being warped, but also history itself. History is rewritten all the time, relative to whatever 'who they are' want you to think, and their political agenda or motivation. And because we have such poor memories, and because we've generally been taught through school to have an unquestionable faith in authority, and not to question anything, and we're generally brainwashed by the media, we accept that what we're being told is the truth, and do not question it.

In fact, I think that our poor memory is one of the things that enables 'them' to keep us disillusioned and enslaved under the system that is in place to entrap us in the financial global regime, and all the nonsense that goes with it.

A Theory of Everything

I'm sure that we're beginning to wake up, though (and pretty quickly), and one of the purposes of this book is to help you in your journey, and anyone else who's interested, and who probably has some knowledge of what I'm talking about already.

I've spent the majority of my life endeavouring to find myself in a position of having a better memory, including engaging in many courses and reading much literature on the subject. People with better memories, for whatever reason, appear more akin to being in the category of 'genius' than others. I'm pretty sure that this has something to do with it, but I accept that I'm speculating, without any hard evidence to present this as being the truth.

For example, I studied Jim Kwik's 'Change your brain, change your life', but I'll never forget once investing in a course in my early days, which was Dr Bruno Fürst's 'Memory and Concentration Studies'. After a week of learning to construct a wardrobe with ten hangers in my mind, and learning ten respective association triggers for each coat hanger, I was then successfully able to remember ten 'things' in any order, at any time, which was a great start. I was excited and encouraged by this; however, I then found it to be somewhat of an oxymoron, that in the next stage it informed me that I *then* had to *remember* 100 further associated coat hanger triggers, and whether rightly or wrongly, I subsequently binned the whole exercise as a complete waste of time!

From the research that I've conducted, it isn't surprising that our memories are generally this bad, though. There's a school of thought backed by scientific research that suggests that a long time ago our second chromosome was manipulated to make it less effective. This is the one that's responsible for memory. Here is a short text by Michael Tellinger:

Reference: Chromosome Fusion (tinyurl.com/y8r3ejru)

And a Gregg Braden documentary:

Documentary Reference: GREGG BRADEN: The TRUTH About EVOLUTION and Your DNA & What it Means for YOU! | Human by Design (youtu.be/OZz_ssh7f4w)

In combination with this, it would appear that early on in our lives during our lengthy time at primary and secondary school, our memories are filled with predominantly unimportant facts and other nonsense, relative to how to live harmoniously together.

In general, what we learn in such institutions isn't necessarily going to help us form a real version of ourselves, our current reality, or our past history. Allow me to elaborate.

1.9 What They Don't Teach You in School

I have nothing but the utmost of respect for the majority of teachers, and people in the teaching profession. From what I've seen of the staff at Alex's school, for example, they blew me away with their dedication, conviction and ethics. For example, trying to deal with his so-called 'ADHD' condition, which has been suggested may be purely a symptom of a globally broken schooling system (see animation link below) and their attention to core life skills, like mindfulness. And also, on the face of it, how they appeared to be trying to circumnavigate the apparent global incentive by actually trying to teach you to think.

RSAnimate – Sir Ken Robinson – Changing Paradigms (tinyurl.com/46jnvhv2)

As far as 'ADHD' and any other labels go, to quote Peter Breggin MD and Psychiatrist, "There is no evidence that ADHD is a brain disease or even a medical disorder. It is not associated with any biochemical imbalances. Studies that supposedly show brain shrinkage in children diagnosed with ADHD are in fact showing stimulant drug damage. The symptoms are nothing more than a list of items that require extra attention from parents and especially from teachers." He has fought tirelessly in court cases and won. He is highly respected in his field and with educating others about the real truth, not only with ADHD but the whole area of medicine and psychiatry. His books and website are brilliant. See for example:

Book Reference: *Your Drug May Be Your Problem, Revised Edition: How and Why to Stop Taking Psychiatric Medications* by MD Peter Breggin (tinyurl.com/myrun6ew)
Reference: Peter Breggin (breggin.com)

Breggin also explains how psychiatry labelled all of these ailments to create drugs to go with them. A calculated collaboration with the pharmaceutical companies, of course.

Instead of creating more labels such as ADHD, a better approach would be to look at the root cause of a child's behaviour. This could even be dietary or environmental. Things like sugar and food additives can play a massive role in behaviour. I've seen children go from quiet and normal to rockets gone off course in the time span of under 30 minutes after having an unsuitable meal, drink, or snack. And let's not forget that some kids are simply the moving, creative type, not the sit-down-and-be-quiet type. They learn and function better while moving instead of spending the day in an almost static position within the confinements of the average classroom.

Whether a child has 'ADHD' or not, I have come to the conclusion that mainstream schooling doesn't appear to teach us or our children much of anything that is of any practical use in life, other than perhaps

Maths, English (grammar, reading and writing) and Science. This is a well-documented topic.

In general, these are some of the key elements of life that mainstream schooling does not focus on teaching us:

- What life's really all about, and that much of our taught and socially accepted understanding of the world and how the universe works is based on 'theory' only
- The importance of improving your own internal dialogue, as well as your communication with others, for example, the study of NLP, which has essentially been developed as a marketing tool, but if you read between the lines you can begin to understand how the specific words that we use influence our thoughts and behaviour
- Emotional awareness and intelligence, or wisdom
- How to keep your body healthy and the importance of a healthy diet (some schools around the world are addressing this)
- How to be a parent
- Professional etiquette and manners (some schools around the world are addressing this)
- The importance of travel, and putting yourself outside of your comfort zone
- How to handle money or your finances
- About love or relationships
- How to run a business, or how to build a career that's all your own
- About the government
- About history
- How to focus
- How to survive without technology
- How to survive in nature, grow or forage food, make a fire and how to purify water

A Theory of Everything

- The dangers of watching TV and of believing everything you watch or see in the media (some schools around the world are addressing this and teaching mass media education akin to critical and analytical thinking)
- About the pitfalls of credit and debt and the truth about money and banks
- How to buy a home or car, and how to do repairs
- About the true dangers of alcohol and addictions
- About healthcare and the benefits of natural healing remedies and spiritual healing practices versus the well-documented, often poisonous and prolonging effects of man-made medicines
- About the Bible, Koran, the Vedas or any other sacred scripts (some schools around the world are addressing this, and also from a spiritual perspective. Spirituality is not a religion, it is philosophy for life!)
- Self-defence and first aid
- Learning from 'failure', and the importance of doing so
- The law, and how to operate in society effectively, but without compromising your life, or making things difficult for yourself
- Finding a job, or time management
- How to avoid (mental) breakdowns or burning out
- How to bounce back, or the art of falling with grace
- How to go through your teenage years, puberty, and what to expect (some private schools around the world do have a guidance program that addresses this; however, they are often only accessible to a few financially rich people, who can afford exclusive, private schools)
- How to cope with problems related to mental health
- How to network effectively (it's not so much about what you know, but who you know)
- How to use your instinct or trust your gut

- Any of the core life principles for harmony and happiness
- The amazing capability and significant power of the human mind and how to think
- About alchemy
- About the occult, Freemasons and secret societies
- The truth about so-called 'drugs', narcotics and psychedelics
- The significance and power of our pineal gland
- The significance and power of sound energy, and magnetism
- The significance and importance of the toroid and its occurrence in universal energy
- The significance of the numbers 3, 6 and 9 – the keys to the universe (Nikola Tesla)
- How to believe in and follow your dreams

I have elaborated on many of these points in Appendix 3. I think you'll find it useful to read through my thoughts.

1.10 And You Have to Go to School

Learning to read and write, and learning about Maths, Science and English could be viewed as being of exceptional importance to your healthy development and future, and I advise you to pay close attention to these subjects as you grow up. Also, we still live in a 'diploma or degree' world. Your qualifications will kick start your life and 'career' at this time. For this reason, I took on a multitude of seemingly random jobs to enable me to attend university and to finish my doctorate degree.

But the real truth is, you simply **have to** go to school. I'll say it again. You *have to* go to school.

Mainly because otherwise, it's possible that your Mother and I would have to go to prison. Isn't that interesting? Yet another story entirely.

A Theory of Everything

Under the circumstances, that option really wouldn't be good for you or either of us! It's just one of those things...

Equally, without taking radical steps, we **have to** live in society, and we **have to at least be seen** to conform, and we currently live in a society that is heavily influenced by formal education, and a 'good' education will give you a good head start in your life and career at this time. This start will enable you to more easily derive your future career around doing things that you enjoy, and that you have an instinctive calling for.

As you grow up, many times in your life you'll have to do things that you don't necessarily want to. However, you can take responsibility for your own experience and make whatever it is that you have to do, as fun as you can, within reason, and within the boundaries that are set.

For example, I did pretty much every manual job going on the temporary work force front during my time as a student. Most of the jobs I did, I'd have been very well positioned not to have enjoyed at all. I've worked in a milk factory, cleaned schools and colleges, been a garage attendant, been a baker, a postman, worked in a garden machinery shop, worked in electronics, worked in a patent office, had paper rounds, worked in data entry, been a dustman... and the list continues.

I wouldn't have personally classed many of those jobs as enjoyable. But what I found was that it was **my choice** as to whether I enjoyed them or not, and to what degree I enjoyed them. Within the confines of the rules, and with respect to others (mostly), I always challenged myself to find a way of enjoying them. For me that was often from the personal kudos I gained from getting better at what I was doing, or coming up with a way of improving the process or task that I was doing, for example.

I think this was the primary reason that I found myself as 'top temp of the year' for two years in a row at Kelly Services in the town where I lived thanks to my very good friend, Mo Spires. I hadn't done anything special other than take great pride in my work, no matter what it was, and tried to be my best, and make it fun for myself and the people around me (with varying degrees of success).

Grandma always told me that "if a job's worth doing, it's worth doing properly". She also taught me that "The more you put into life, the more you get out of it".

NEAT LIFE HACKS: YOU'VE REALLY GOT NOTHING TO LOSE, AND EVERYTHING TO GAIN FROM TRYING TO ENJOY BEING AT SCHOOL, OR WORKING TO GAIN THE NECESSARY AND APPROPRIATE LIFE SKILLS AND EXPERIENCE

But going back to the schooling, for either of you; please don't ever let the system break your spirit. Because being the way you are, I'm sure will give you a head start in life, once the schooling is behind you, and you can begin to educate yourself in what it is that's important, what interests you, and for the rest of your life, the ability to follow your dreams!

NEAT LIFE HACKS: ACCEPT THE FACT THAT YOU UNDERSTAND THAT THE PRIMARY PURPOSE OF SCHOOL IS TO TEACH YOU TO CONFORM, AND TO ERODE YOUR ABILITY TO THINK CRITICALLY, AND TO BRAINWASH YOU TO FIT INTO OUR BASTARDISED SOCIETY. ACCEPT THIS, BUT WITH THIS KNOWLEDGE IN MIND, PAY ATTENTION TO WHAT'S IMPORTANT TO YOU, AND WHAT INTERESTS YOU, AND PRETEND VERY HARD THAT YOU ENJOY ANYTHING ELSE, NO MATTER HOW TEDIOUS YOU FIND IT, PARTICULARLY PAYING ATTENTION TO NOT DISTURBING THE 'LEARNING' OF OTHER STUDENTS. YOUR TEACHERS IN GENERAL ARE TRYING TO HELP YOU, AND WANT TO HELP YOU IN THE BEST WAY THAT THEY CAN WITHIN THE SYSTEM THEY HAVE TO WORK WITH. TRY TO MAKE IT EASIER FOR THEM, AND ABOVE EVERYTHING, TRY TO MAKE WHAT YOU DO ENJOYABLE AND FUN FOR YOURSELF (WITHOUT DISRUPTING THE REST OF THE CLASS!)

A Theory of Everything

1.10.1 ALEX, YOU'RE A MINI-ME!

Actually, Alex, you're so much like me when I was a child in school, you're almost a mini-me! I didn't mean to display behaviour that upset my classmates and teachers, and I didn't understand back then why I was always in so much trouble either. And not just as a child in school, but also at home and in my community. I certainly had back then, and still do to a degree today, have the classic symptoms used to diagnose and label 'ADHD'.

One of the main differences between our generations, other than the new global outlook on so-called 'ADHD', and other spectrum labelling, is that I didn't have all this electronic distraction, or even much TV for the majority of my childhood and that the worst form of communication ever conceived, texting, didn't exist yet.

And as a result of my accident when I was eight years old, I made it my personal mission to prove to myself, if to nobody else, that, even though I'd suffered 'brain damage', I was still clever, and I could still exceed in every subject they threw at me, no matter how boring or irrelevant it was! This obsession led me to the highest level of academic qualification, completing my Doctorate at the University of Oxford, and although I've often looked back and wondered how different my life would have been if I had not taken this path, I have no regrets.

Equally, I didn't have as much exposure to the constant bombardment of misinformation, propaganda, and influence such as advertising, and, 'you have to buy our product, or the puppy dies' messages that are now prolific in TV and radio in today's Western society. I'm sorry, but the puppy usually dies.

A Theory of Everything

NEAT LIFE HACKS: UNDERSTAND THAT THE PRIMARY PURPOSE OF THE MEDIA, NEWS AND TELEVISION IS TO PROGRAM OUR MINDS AND INSTITUTIONALISE US INTO BECOMING THE SLAVES THAT, FOR THE MAJORITY, WE ARE. FREE YOURSELF AND YOUR MIND BY QUESTIONING EVERYTHING YOU READ OR HEAR, PARTICULARLY IF YOU RECOGNISE THE MESSAGE AS BEING NEGATIVE, AND REJECT ANYTHING THAT DOESN'T FEEL GOOD, OR RESONATE WITH YOUR CORE. TAP INTO YOUR GUT FEELING AND INTUITION WHENEVER YOU CAN

Also, as a result of no mobile phones, computers, or TV as we know it today, I spent the majority of my childhood outside playing football (not as well as William, of course), playing tag with my friends, building dens and fires, camping, sailing, fishing, climbing trees and seeing who could jump from the highest branch, blowing stuff up, and generally getting into more trouble, as I'm sure Grandma would be happy to fill you in on! In this respect, I'd pretty much already burned up most of that pent-up 'ADHD'-fuelled energy and frustration, before I even interacted with anyone who expected me to be 'good'. And I'd done this no matter what the weather was doing, or where we were.

We didn't have access to many sweets either, which in hindsight was an exceptionally good situation! I feel for you, Alex. I really do. Because I understand the allurement of sweets, but I also understand that commercially available sugar is a poison and is exceptionally bad for you. It's thought to be responsible for, among other dietary and physiological things, damaging your brain function especially as you age. I'm pretty sure that my brain's been somewhat addled by the amount of sugar that I subsequently ate later in my childhood and early adulthood, and the sugar content of the alcohol that I consumed later in life.

You see, the body naturally craves glucose, a ready-to-use form of energy for our brain and muscles found in sugary foods. And it can be found in fruits and other natural products. But what appears to have happened is that the food industry has capitalised on this natural craving and sold us refined sugar cane as an inexpensive and unnecessary substitute. This stuff really messes with our minds and bodies. Alex, did you notice that at the same time you are more inclined to eat sweets and sugar than William, you are also more inclined to eat fruit. I am sure that there is no coincidence!

Oh, and I wasn't constantly subject to having my picture taken all the time either! That must really drive you nuts! It would be interesting to understand what the psychological effect of this phenomenon has on the youth of today, compared to people of my generation.

As I said, although I've learned to change my behaviour to make myself more effective in pursuing my goals in general, I am, and always have been, 'ADHD' positive. If it had been diagnosable when I was a child, I would have been undoubtedly on the 'ADHD spectrum'. The greatest value I took from being like that was my utter distaste in being told what to do, and my utter disrespect for most authority.

I get bored very easily and find it difficult to concentrate on anything that doesn't interest me. In meetings at work, for example, I find that playing solitaire in the background actually helped me to concentrate on mundane or uninteresting things which I know I have to listen to, for whatever reason. Driving was challenging when I first started to learn. I had 12 accidents in the first two years of my driving career, before I worked out that I had to concentrate at **ALL** times, and not *just* when I was driving fast. Even today, I struggle with patience and consideration for others, although my core value is really not to upset other people.

To this day, as you are already aware, one of the most empowering and concisely structured messages, which I have lived by, is contained within the lyrics of the song *Killing in the Name* by 'Rage Against the Machine'!

Don't live by these lyrics though. Simply do your best at school and

A Theory of Everything

at least pretend to conform, but do even better outside of the classroom. Do your research, and educate yourselves in the truths that are gradually unfolding in front of you.

"A child educated only at school is an uneducated child" George Santayana

"The school system is designed to teach obedience and conformity and prevent the child's natural capacities from developing" Noam Chomsky

"Education is what remains after one has forgotten what one has learned in school" Albert Einstein

---

Part 2:
My Behaviour, and What I
Wanted to Change

2.1 My Own Behaviour as a Child

Let me explain to you some of the challenges that I faced as I grew up, and what led me to my current understanding of human psychology, life, the Universe, and everything as I know it.

When I was a child, I was absolutely guilty, for whatever reason, of bending the truth to make things I'd seen or done sound better than they actually were. I was equally guilty of making excuses when I didn't want to do something, or if I had done something wrong. I know where that trait came from.

I did then and, to a degree, still do to this day, feel the need to impress people with who I am, even though I know that I really don't need to. I think it's a hangover of feeling that I was so misunderstood as a child with physical limitations and 'ADHD', and wanting to show to the world who I really was.

For most of my younger life, I was also very guilty of assuming that I knew something, because I'd heard it from *whatever* source, or even just pretended that I knew something (winging it), and thinking that I could talk with authority on whatever topic it was.

I grew out of this in time, but I learned the hard way and spent much time trying to work out how to change this behaviour. I saw that, for as good as I was at gaining trust in the early stages of a relationship, when I was eventually found out to be a fraud, then the relationships were destroyed, often irreparably.

And I realised that there was no need for this at all, and that it was absolutely not how I wanted to behave, or how I wanted to come across. I realised that I could have those relationships, and be as honest and authentic as I wanted, which I desperately wanted to be. And actually, if I was completely honest in my communication, then the trust became harmonic, and I could develop strong and positive long-lasting relationships.

I became acutely aware of the fact that who I was would transpire in my words and actions if I was simply myself, in my most authentic presence. And in that respect, I didn't need to orchestrate the verbal diarrhoea that I had always been prone to, and my apparent need to talk about myself. And that if someone didn't like me the way I was, well, then that was indicative of the fact that they were not vibrating at the same frequency as I was, and were simply not meant to be in my life then or ever, or me in theirs. I could accept it and move on.

So, if nothing else, I urge you to think about the information you're going to convey to someone. Always ask yourself (relative to the relationship) whether you know for a fact that the information is true, or, at the very least, make it clear where you got your information from, and how confident you are in its source and authenticity, and your own authenticity. And more than anything, particularly if it's something you're interested in, do the research to put yourself in a better position of knowledge and authority the next time the topic is discussed. Don't talk to impress, often less is more.

William, I'm extremely proud of you, because you appear to get this already, and I'm sure that it will make your life more positive, harmonious, simple and productive as a result!

"Cheers to all of those who do their own research. In the age of information, ignorance is a choice" Leonardo DiCaprio

---

NEAT LIFE HACKS: ALWAYS RESEARCH ANYTHING THAT YOU WANT TO KNOW ANYTHING ABOUT, PARTICULARLY THINGS THAT YOU'RE INTERESTED IN AND NEVER TELL ANYONE ANYTHING WITH CONFIDENCE, IF YOU ARE NOT HONESTLY CONFIDENT IN WHAT YOU ARE SAYING YOURSELF – PEOPLE IMPORTANT TO YOUR LIFE AND DEVELOPMENT WILL RESPECT YOU FAR MORE THIS WAY, AND WILL BE MORE INCLINED TO LISTEN WITH INTEREST TO ANYTHING YOU DO HAVE TO SAY. GENERALLY, I'D ADVISE THAT IF YOU DON'T HAVE ANYTHING INTERESTING OR INFORMATIVE TO SAY, THEN DON'T SAY IT (WHICH IS RICH COMING FROM ME, HAVING BEEN A PART-TIME GOBSHITE FOR MOST OF MY LIFE)!

If you think that I've violated this rule at any point in this book, then I would like to know, and I want to make the necessary reflection and correction! My intention is to practise what I preach.

2.2 My Behaviour as a Young Adult

When I was in my early to mid-teens, I became even more frustrated with my behaviour. I wasn't happy because still the way that I thought I should, and wanted to, behave was not reflected in my actions.

I was acutely *aware* of my behaviour, as William seems to be of his, and I couldn't understand why I behaved in a way that I was not happy with, even though I routinely told myself how I wanted to behave, and then beat myself up regularly for not being able to change.

It seemed pretty obvious to me back then. I knew how I wanted to behave, and what I wanted to change, so why couldn't I just change it? I mean, what's that all about?

The reason for this didn't become clear until I was in my mid-thirties, and I'll explain that a bit later on (In Part 6).

2.3 Fundamental Human Psychology

So, what I did back then, in my early to mid-teens, was to study the behaviour and psychology of other people.

I did this from scratch. I just observed people that I knew. I didn't read any books in particular, and I didn't consult anyone for advice, other than what I gleaned from various counselling sessions. I just did it gradually, and learned along the way, to the degree that, right now, I feel that I have a healthy, albeit academically unrecognised, understanding of basic human psychology, and beyond.

This is a topic that has fascinated me ever since, and even during my formal education, if I hadn't gone down the path of Energy, Science and Engineering, I would certainly have gone down the path of Psychology. I have no regrets!

So, what was I doing studying other people? How was that going to help me?

Well, I figured that if I could begin to understand what it was that caused other people to behave the way that they did, then I may eventually be able to superimpose that understanding onto myself, and to work out exactly why I behaved the way that I did. And I might finally work out how I could change my behaviour, without this continual beating myself up for behaving like what I perceived to be a lot of an idiot.

At least, that was my theory.

This became a lifelong quest and obsession. I gradually learned, step by step and from first principles, why we have the core beliefs and ideologies that we have, and subsequent behaviour that we display, and why it's so difficult to change these behaviours, even if we're **aware** of them and desperately want to change them. I even learned about mental health issues such as addictions or depression, and where I thought they stem from, and had many ideas on how I thought they could be effectively treated. What I learned changed my life forever, and I believe has contributed to the happiness I've had over the last several years, despite the roller coaster ride that I persistently expose myself to.

NEAT LIFE HACKS: IT ALL BEGINS WITH BELLIGERENT AND BRUTAL SELF-HONESTY IN BECOMING AWARE OF YOUR BEHAVIOUR, AND ALLOWING IT AND ACCEPTING IT, SO TO LAY THE FOUNDATIONS FOR THE TASK OF THE REBUILDING OF THE TRUE YOU, AND YOUR ULTIMATE SUCCESS

You see, what I learned was that it's a lot to do with our 'preprogramming', or the memories that evoke feelings or emotions, which are embedded deep into our subconscious. And just to demonstrate how challenging it is to access and reprogramme your deep subconscious, a Buddhist monk would probably have to spend 20 years sitting on a mountaintop 'Ohmming' to eventually reach the state of deep meditation, where his subconscious lies, and becomes accessible and workable, to remove any of the unwanted beliefs and preprogrammed junk. On a side note, I learned that there's a lot more to 'Ohmming' than is conventionally understood, and the significance of **the power of sound and vibrational frequency**, and their influence on life and creation.

But I didn't have to spend 20 years sitting cross-legged on a mountain contemplating…well, absolutely nothing really, as that's one of the basic principles of meditation (see section 9.11). Instead, I found that it was made far easier for me by taking advantage of new technology, and by combining various techniques to get closer to the results that I wanted. I investigated and experimented with a plethora of techniques including Emotional Freedom Techniques (EFT), meditation, binaural beats, and other energy and psychological techniques, which I will describe to you further in this book.

As I've mentioned before, I'm acutely aware of the fact that I'm not completely 'there' yet, wherever 'there' is. And, I'm equally aware of the fact that I may even have a long way to go. But if I look back at what I feel I've learned and achieved, and the tangible results that I have seen, then it is clear that I've made significant progress, and have come a long way already. And surprisingly enough, one of the greatest challenges in my life, my addiction to alcohol, helped me greatly along this journey.

2.4 My Career with Alcohol and the Foundations of the 20-Year Road to Recovery

I'm painfully aware of the psychological damage I must have caused you when you were younger, due to my drinking and my anger.

I'm not going to beat myself up. Back then, I was doing my best with the knowledge, information and capabilities that I had. But I know that I made mistakes. And I often let my drinking and my 'anger' tarnish the great father that in my mind I thought I was, could be, and that I wanted to be.

Yeah, I had a plethora of excuses for my drinking habits, and I thought I had better excuses than most people, with my forever present physical pain, paired with the insomnia I'd had since I was a child. But they were just that: excuses. The pain and insomnia were real, but anger and drinking weren't the solution. And it wasn't until I could openly and honestly accept that they were excuses, that I could properly begin to address my addictions, the beliefs causing and stemming from those addictions, my thoughts, emotions and subsequent behaviour.

A Theory of Everything

Eventually I even learned where those addictions, behaviours and addictive personalities came from in the first place, and I learned how to go back and heal the past, in this lifetime, and in past lifetimes too.

But let's go back to the beginning. How did my career with alcohol start?

See, I always thought that I could learn to control my drinking. I was convinced that one day I could learn to moderate my alcohol intake, and to be able to drink like a 'normal' person. I thought I enjoyed it, and that I got benefit from it. I'd stopped on numerous occasions and had very strong will power, if I wanted it. I thought that if I was determined enough, I could. I knew that it was in my nature to be an all-or-nothing guy, and I assumed that this was the single reason for why I was struggling to moderate my alcohol intake. I'd already acknowledged long before that I was an alcoholic. That wasn't the problem. The solution was in understanding what being an alcoholic *actually* meant to me.

I must have been in my mid- to late-teens when I took my first drink. I remember it very well.

I'd undergone several operations to tendons in my right lower limb in order to attempt to correct the length deficiency caused by stunted growth at the age of eight when the accident occurred. The operations spanned between the ages of twelve, when I suffered a tilted pelvis, to the age of nineteen when I had tendon extensors on the four toes of my right foot. Each of the surgeries I had were pioneering efforts by my young and enthusiastic orthopaedic paediatric consultant. Like, for example, when I had my groin tendon extended, he wrapped a plaster around my thigh and one around my waist, so tight that I struggled to breathe (and for a period could not actually breathe deeply), and connected the two with a Boat Joyce. A Boat Joyce is a rod in three sections where, when you turn the middle section the whole rod either becomes longer or shorter depending upon which way you turn it. It was used so that every morning I could turn the middle section a couple of times, and thus gradually extend my tendon without too much discomfort. Unfortunately, the contraption failed during a holiday in North Wales and we ended up at a boatyard for advice. All the doctors at the surgery and hospital had no idea what to do! Then my mother (Grandma) held me down while my nannie (your great-grandma) drilled into the wood in my plaster in order to repair the device. I found this terribly amusing!

That was it, though, at the age of nineteen I said, 'Stuff this', the operation on my toes had not been successful and had left me with so much pain and aching that I started to drink heavily as a result, and I just wanted my life back.

And I didn't fuck about either. My first-ever proper drink was a bottle of Bell's whisky! In the same respect that the first thing I ever smoked was a joint, my first drinking session was hardcore spirits. I was at home in Horsham, where I lived with my mother (Grandma), stepfather (Grandpa), sister and stepbrother, and I remember that I'd just been bought a new bed, and the first night that I slept in it I vomited a rancid pool of alcohol-infused stomach content onto it. I was depressed, and I am pretty sure that my 'attempted suicide' was nothing more than a cry for help. Grandma didn't say a word about it. Yes, I knew I was in the shit... but she didn't say a thing, and I spent the rest of the day in her bed recovering.

Moving forwards, I'd binge drink, like most of my peers at that age, but it stopped briefly once I passed my driving test and preferred to drive rather than drink. I also smoked dope instead of drinking, and since dope-driving laws were not as well established as drink-driving law enforcement, it seemed to be the better option. I'd tried drink-driving and it wasn't for me!

For at least a couple of years I had good incentive not to drink on such a regular basis, but when I got to university this all changed. There were two main factors that I allowed myself to use as an excuse for drinking back then, the first being insomnia. I just couldn't switch my head off at night, and alcohol allowed me to do that, and allowed me to sleep. And frankly, the ramifications of alcohol in the morning were no worse than those of the sleeping pills and painkillers that I'd been prescribed. So, when at university I consulted the GP for a solution to the pain that I was still enduring from all the operations, and when the painkillers that they gave me made me feel terrible in the morning, it was easy to justify alcohol as self-medication. At least with alcohol I was able to derive some perceived pleasure from administration the night before! But I couldn't see that I was an alcoholic back then. I couldn't see that I was drinking every night for any other reason than that which I have described. I remember that my local corner shop owner said that he was concerned about me, and I remember being curious and actually grateful, but ultimately ignoring him, thinking I knew better, and that I knew what I was doing. I absolutely did not know better, or know what I was doing!

I remember that as students we'd go out every Monday night, which was students' night, and drink at various establishments en route to the club, until I was sick; then I'd continue drinking until I couldn't remember how I got home. But I always did get home. And I remember that one of my party tricks was that I could fall over from the bar with a full pint in my hand, and not spill a drop. And I could drink a pint and then proceed to expel much of it back into my pint glass through my nose. These achievements for some reason gave me great pride, and added to my passion for drinking. The one thing that I couldn't do, however, was to prevent myself from needing to take a piss all the time once I'd started drinking. In time, this led to stronger and stronger alcohol-content drinks, to reduce the necessary quantity of fluid that I had to consume (and pass), in order to get me to where I wanted to be. And pub crawls were not my thing. With limited mobility and consistent pain, it drove me nuts having to walk so much and having to stand in a pub if there was no available seating.

A Theory of Everything

I remember that when I did my internship in Žilina, Slovakia, I was known as 'Jonney Red Eyes'. On several occasions, after a heavy night drinking in town, I remember waking up in the late morning on the path to the department having had a lie-down on the way because I was so hung-over. It didn't occur to me at the time that I was damaging my career prospects, because, when I was on it and in the zone, I was on fire and achieved great things; so I allowed myself my tardiness on the odd occasion. On more than one occasion I arrived at the department so late that everyone else had already gone home for the day. So I went back out on the town again.

I did my first degree at the University of Newcastle-upon-Tyne, which is where I started my second degree, but it wasn't until I transferred to the University of Oxford towards the end of this time that I began to be aware for the first time that I had a problem with alcohol. I remember that I stopped drinking for a period during this time, and sought medication for sleeping. My doctor at the time recommended that I stop alcohol but continue smoking, and he prescribed me temazepam for sleeping. My God, it was good! I remember the feeling of having had a proper night's sleep for what seemed to be the first time in my life! I actually turned up for the 08:30 group meeting for the first time since I'd been part of the group at Oxford, and I think I even attended the one the following week, too! But this was short-lived and a slippery slope back to alcohol was to ensue.

By the time that I had graduated and found a partner who was to become my wife and your mother, I was drinking heavily and had graduated to drinking predominantly wine and spirits. I was drinking to get drunk, and for no other reason, and was unsatisfied if the results of my endeavours ended any differently. And my reasons for drinking harder alcohol were simple and effective. As I mentioned, if I drank beer or lager, rather than wine or spirits, then the frequency for me to need the toilet was greatly increased, which was annoying. Equally, and put quite simply, wine and spirits got me to where I wanted to be much quicker.

On top of that I had significant mood issues, which I became more acutely aware of over the following decades, but in the early days I allowed them to contribute to the plethora of subsidiary excuses for drinking excessively. I'm an emotional guy, and back then I wasn't able to control my emotions, and I wasn't usually afraid to show them either (as both of you well know). Back then, I was also still strongly psychologically affected by my childhood accident and the physical restrictions and pain that it caused. It took me decades to come to terms with this. Drinking seemed to ease the thoughts and pain.

I was immersed in a drinking culture, particularly from the student background that I'd enjoyed for so long. I mean, for a period when I was lecturing at Newcastle, I would take my students to the pub for some of the lectures, convinced that they would learn more in this environment; and they never let me down! All of my peers drank, and my highly regarded supervisor, Professor Alexander Korsunsky, was Russian and taught me how to distil vodka from tomatoes and potatoes.

Then, when I got my first job with a large corporation working in the field of fuel cells, the idea of being able to stay in hotels and expenses, my alcohol bill was just too much to turn down! Equally, I found myself within a drinking culture again, and it was normal to justify such expenses. It seemed strange to me that I could, without question, claim expenses for copious amounts of alcohol on a nightly basis, but if I tried to claim for a magazine, questions were asked, and investigations were conducted. I could even claim for room service 2!

My next job saw me in a similar drinking culture but this time in a higher position of authority, and the expenses flowed. I used to travel to Boston quite a bit and remember that the staff in the bar at the London-bound gate at the airport used to know me very well and I was constantly being told in that fantastic American drawl: 'Je-on, youw drink like a fe-ish' – fear of flying was another good excuse that I used to justify getting completely hammered when the opportunity presented itself. I used to fly Upper Class with Virgin quite a lot in my late teens, early twenties (because Grandpa worked for the airline), and they didn't hold back, and so I developed a preference for airlines. Some airlines were far more tolerant to drinking on their flights, and even let you get your own drinks from the galley in economy class, whereas others were far more stuffy, and I think they even refused to serve me on several occasions. I remember on numerous occasions waking up after a flight to find that I had vomited on myself.

So I spent several years of married life in much ignorance and denial about my alcohol problem. Interestingly, both your mum and I stopped drinking and smoking during your conception, for which I'm very grateful to your mother, as well as for her positive influence on me at that time.

After you were born, William, I was frustrated that when I was away from home so much, you didn't recognise me when I returned home. That, along with a number of other reasons, including moving closer and closer towards requiring a wheelchair for mobility, I decided to take matters into my own hands and start my own business coaching consultancy. The idea was that when you, Alex, were born, I'd be working from home and not away so much, and would have more time to spend with my family.

The resulting chaos that ensued, on top of being frustratingly close to both of you and your mum, while simultaneously having less time to spend with you, and the subsequent IVA arrangement and later bankruptcy of a failed business, and the constant letters and calls from creditors that I endured, provided me ample justification for not even considering that drinking was causing a problem and making me insane. Rather I felt it was curing a problem and keeping me sane!

I think that the only times in my life that I've felt suicidal when drunk was when I had significant immediate issues in my relationships. I remember one time after an argument with your mother, I was driving down a hill in the pouring rain at about 70mph and, conscious of the fact that there were no other cars around, I applied the handbrake, slid the car sideways, and crashed into the bank at the side of the road. I was very lucky, and, even though the police actually drove past me when I was at the side of the road, and a passer-by stopped to ask me if I was alright, and must have clearly smelled the alcohol on my breath, I made it home without significant incident and lived to tell the tale with my licence intact. I'd have been absolutely broken if I'd have lost my licence. One other time, and probably when drunk, I began to construct my suicide letter, which gave me some perspective and made me feel better, although I spent a long time contemplating how I would end my life. During my first degree I remember taking an entire bottle of my antidepressants, and spending the night curled up on the floor listening to the bees in my pillow and nervously watching them crawl around my room and ceiling.

It wasn't until my marriage with your mum had got to the point of breakdown for a multitude of irreparable reasons that I was forced into sobriety for the first time properly.

I wanted to know whether by changing my behaviour I could save our relationship and get back to what we had once shared, and whether it would make any difference to the apparent and understandable contempt with which she held me.

For the next two years I abstained from alcohol and researched and practised meditation and self-awareness, and worked hard on myself. I became what I believed to be the same but more wise self that I was when Mum and I had first met; yet after those two years I concluded that, although grateful for the experience, it did not matter how much I had changed, the relationship would never be the same or better again. I started drinking again at Christmas and New Year of 2013/14 whilst with

A Theory of Everything

you at Center Parcs in Rotterdam, and I left our family home between Christmas and New Year of 2014/15. I left everything. I packed a suitcase of clothes and just left, and to all intents and purposes it was the right thing for me to do, and the right thing for both your mother and me, and for you boys, too. By this time, I had a very good job, which I'd created for myself, and was financially stable, and could afford to pay reasonable maintenance costs to maintain payment on half of the mortgage of the house that you lived in with Mum.

But bachelor pad and newfound freedom equalled unrestricted and debauched drinking with no limits, and many a good reason for justification!

I'm thoroughly grateful to the whole experience of the relationship that I had with your mum, and with alcohol at that time, and what I was to gradually learn about myself, and for the self-development work that I was to do on myself. I'm truly sorry for the harm that my alcoholism caused to her and to you, and grateful to her for what she endured, and what she enabled me to understand about myself. It's assisted in bringing me to where I am now, and enabled me to put myself in the best position to develop and sustain better relationships across the board in the future.

During this time, I learned about the deceit and lies that we tell ourselves. I learned about the insanity of addiction. I observed my behaviour and saw the fucked-up and perverse nature of my reasoning, and the efforts that my 'head' would go to in convincing me that I could use any justification available to start to drink, and then carry on drinking until I was drunk. And I saw the interesting tangents between these insanities and, for example, those of the addiction that we often refer to as 'love', when first in a relationship.

Yet still I thought I was in charge – I was very rarely in charge, and once it got a hold of me, I was very far from being in charge! Historically, although I'd shown on several occasions that I was capable of abstaining from alcohol for significantly long periods (up to two years), and that in the immediate aftermath I was able to drink sensibly, it wasn't very long (say, a month or two at the most) before I was back into my old habits of indulgence and insanity. As I said, the first time I properly abstained was forced upon me by the threat of forgoing my relationship with your mother, marriage and family for you boys. The second time, interestingly at Christmas/New Year of 2015/16, was a personal decision motivated by an observation of my deteriorated behaviour, and as a quest to lose the weight that I'd put on during the previous years' drinking in my bachelor pad and various extended business trips.

The third and final time I will describe later in the book, and this time it was different.

2.5 After My Formal Education

At the age of about 25 the student lifestyle came to an end. What naturally happened was that, like anyone else, I had lots of pressure put on me in terms of living *realistically* in 'society'. As I understood it back then, I had to do things like get a wife, have kids, buy a house, pay tax, the water rates, and electricity and gas bills, get *'a fucking big TV'* (which back then was about the same size as my 22" computer screen, and weighed about the same as a small elephant), maintain a car, and everything else quoted in Irvine Welsh's book *Trainspotting*! And I had to get a job and career in order to continue to pay for all of that stuff, and I'd have to work hard and live as harmoniously as possible in society, and act upon all of the other social beliefs hammered into me as I grew up. And I felt I had to do all of this pretty damn quickly, because I was already at least four years behind some of my friends from school and college, who had been 'homeowners' for some time already, and one of them even owned a Lotus!

A Theory of Everything

Jeese Louise, I had a lot to learn about life!

That whole 'having kids' thing changes your life beyond recognition, and beyond anything you or anyone could understand or appreciate from any book you read, or anything anyone can tell you, before you actually have kids yourself. And it did for us, especially as your mum and I had little support from our own parents. Geographically we were simply too far away from them to play a more influential and supportive role.

I don't mind telling you that having you has been the greatest paradox I've ever experienced.

If I had to go back and do it all again, I probably wouldn't do it for *any* incentive that *anyone* could offer me. However, from the moment you were born, I knew I wouldn't change it for *anything* in the world, or even the known universe, and beyond!

If you don't understand this, and many of my younger peers have thought me a little crazy when I've explained it to them, then just wait until you have kids. You'll understand then!

Back then, I was just beginning to learn about the world and the way it worked, and I felt that something big was on the horizon. From my position of relative naivety and fear-programming, the world looked scary and nasty, and it sure didn't look like a world that I wanted to bring other lives into. I didn't want to have children, bringing them into this world as I saw it.

But I'm so glad that my sex drive eventually completely trumped my rational mind, and that I brought you into this world.

"Do you want to have kids?" "No." "Do you want to have lots of sex?" "Yes!"

Because by now, my belief and understanding are that this is possibly one of the most exciting eras of mankind as we know it, and I'm so glad that you are here to experience it, to witness what is going on, and to be a part of it.

"Our ancestors devised every method imaginable to alert us to a single fact; now is the time of the most extraordinary conditions and opportunities that accompany the rarest of events, the shift from one world age to the next" Gregg Braden 'Fractal Time'

"Science suggests that the next step of human evolution will be marked by awareness that we are all independent cells within the super-organism called humanity" Bruce Lipton and Steve Bhaerman, 'Spontaneous Evolution'

"I believe it will be the magnetic influence produced by the sun that will usher in what is described by our ancestors as 'the transition' bringing us to a new state of being" Mitch Battros

---

2.6 Treading Water

All of this trying to conform to society contributed to a slight slowdown in my quest for personal development, but I became more interested again once I was into my second job, and in a more senior position with more responsibility. I remember writing a list of the behaviours and traits that I wasn't happy with, and wanted to change, and how I wanted to behave in the future, and I remember reading it every morning.

A couple of examples on this list were my inherent shyness, and my occasionally uncontrollable anger. I wanted to become more calm and confident, yet remain assertive, and in control of my emotions. I didn't quite know how to tackle it all, but I remember starting the slow process of researching different methods to help me with this.

It wasn't until my brief business sabbatical that I was to truly learn the significance of self-development. I distinctly remember the day that it all clicked for me. Sure, it's not rocket science, but I realised what I needed to do in order to start the journey towards what I wanted to achieve. What I needed to do was to take *full* responsibility for myself. Both in my personal life, and in my working life.

You see, in my professional life I'd always had the attitude and mentally that, if a particular employer wanted me to learn something new, then it was *their* duty to provide *me* with that knowledge and information. In their time, and at their cost.

I was wrong… So wrong! Epic fail!!!

A fundamental mistake. I suddenly understood that if I wanted to get better at anything, be it my own behaviour, or my professional skills, I had to take as much responsibility as possible, and invest my own blood, sweat and tears in anything that I wanted to improve. This realisation was the beginning of an incredible journey of self-development, acquisition of knowledge and enlightenment.

NEAT LIFE HACKS: TAKE RESPONSIBILITY FOR YOUR OWN SELF-EDUCATION AND LEARNING, AND ACTIVELY RESEARCH THE THINGS THAT INTEREST YOU AND EXCITE YOU, OR THINGS THAT YOU NEED TO UNDERSTAND BETTER TO COMPLETE A PARTICULAR TASK, GOAL OR YOUR ROLE AT WORK. MAKE BEST USE OF YOUR OWN TIME TO DO THIS. YOU CAN ONLY BENEFIT FROM THIS YOURSELF IN THE LONG TERM, EVEN IF YOU THINK THAT YOU ARE DOING IT TO BENEFIT SOMEONE ELSE IN THE SHORT TERM

"Never wish life were easier, wish that you were better" Jim Rohn

2.7 Fear and Risk

One of the first things I tackled was my inherent shyness and fear. It was one of the things I'd always found very challenging as a child, and while growing up. And it completely baffled me. I was trying to understand why

A Theory of Everything

I was so inherently shy, and reserved, and scared to take opportunities that were presented to me. I spent a great many years in *regret* for the things that I *hadn't* done, and the opportunities I'd clearly missed.

I finally started to consciously put myself outside of my comfort zone, and to take the opportunities which came on my path. I realised already back then that fear like this, closely associated with negative memories stored in your subconscious, was your enemy, and, in that respect, was pure 'evil'. I think this is how I subsequently developed such *'big balls'* in my classical all-or-nothing approach to most challenges which followed.

Don't get me wrong, fear can be an invaluable tool, in telling you what's dangerous and life-threatening. But you see, I think there's a limit to what this can teach you, and beyond that, it's far more of a hindrance than a help, and influenced by everything in your environment around you. That is, until you recognise what's happening and take steps to *control it*, rather than *it controlling you*.

"Close your eyes and let the mind expand. Let no fear of death or darkness arrest its course. Allow the mind to merge with Mind. Let it flow out upon the great curve of consciousness. Let it sour on the wings of the great bird of duration, up to the very Circle of Eternity" Hermes Trismegistus

---

NEAT LIFE HACK: ALWAYS STRIVE TO PUT YOURSELF OUTSIDE OF YOUR COMFORT ZONE, WITHIN REASON. WHAT YOU'LL FIND WHEN YOU PUT YOURSELF OUTSIDE OF YOUR COMFORT ZONE IS THAT WHEN YOU LOOK BACK ON WHAT YOU'VE ACCOMPLISHED, YOU CAN SEE HOW MUCH EASIER IT LOOKS FROM THE OTHER SIDE, AND THAT THE FEAR THAT YOU EXPERIENCED PRIOR TO IT HAS EVAPORATED AND BECOME INSIGNIFICANT

And that brings about the question of **risk**. Now that's an interesting one. I became more risk-conscious.

If you remove unnecessary fear, then you're left with risk management. Left to assess the possible dangers and ramifications, potential benefits or losses, and how quickly you should take action, from putting yourself out there and taking those risks.

Somehow, I seemed to have been able to assess them, with a little caution, but not much, in a way that I'm still alive today, although your grandma may be inclined to disagree. Not to disagree that I'm still alive, though, you understand!

"Being 'realistic' is the most commonly trodden road to mediocrity" Will Smith

---

I believe that what Will's referring to here is being *'socially'* realistic. I find that there's a big difference in realism once you strip away all the social lies, misinformation and preprogramming.

"Life's not a spectator sport. If watching is all you're going to do, then you're going to watch life go by without you" The Hunchback of Notre Dame

---

2.8 Addressing my Anger

Another element of my behaviour that I found very challenging was controlling my anger. And this was particularly the case when I'd been drinking.

What challenged me the most about it in the early days was that I understood from various sources that a direct result of a head trauma could be excessive anger. I wasn't put off by this, but I also by then had realised that for my own sake, I couldn't use this as an excuse to make it OK.

I attended various anger management courses and seminars, and I read many books and articles in my quest for knowledge and information. One of the striking things I learned was that most of our anger stems directly from negative memories, which are often fear-based, and that are locked deep in our subconscious, and that is why they are so difficult to work with and change

I've described how I've addressed anger later on in this text, and detailed the methods that I've successfully used to observe, balance and harmonise my emotions. I've described the process that I developed to remove and change the negative memories, or blockage points, in order to transform yourself into the person you want to become, and who you truly are.

At this stage, though, I'd say that I've personally found the chapter on Anger in the book *Letting Go: The Pathway of Surrender* very valuable. Written by David R. Hawkins

Book Reference: Letting Go: The Pathway of Surrender by David R. Hawkins MD PhD (tinyurl.com/y6veukl6)

2.9 Quantum Connection

Throughout the years, I found numerous other techniques for my emotional development, most of which were in direct conflict with my trained 'classical physics' scientific approach to, and my understanding of, the world. But the more I learned, the more I began to accept a different way of looking at the world.

And to my absolute delight, it appeared to be directly in line with my passion for understanding energy and resulting fascination with quantum physics. I remind you of this quote again:

A Theory of Everything

"If you want to find the secrets of the Universe, think in terms of energy, frequency and vibration" Nikola Tesla

---

What I learned was that many of the commonly observed 'unnatural' phenomenon that kept popping up in my research more and more, such as energy healing, remote healing (both assisted with scientific equipment, and unassisted), paranormal activity, clairvoyant abilities, telepathy and the significant power of the untapped mind, and similarly socially pooh-poohed claims from perceived whacked-out, crazy people, could actually be explained scientifically and from a quantum perspective, thus making these claims very much fathomable, tangible, and understandable, and the respective practitioners (ignoring the charlatans) quite sane in their work, ambitions and abilities!

Part 3
The Turning Point

3.1 Anything Is Possible

Around about the time that both of you'd been born, I began to develop and quantify my understanding that anything is possible, and that anybody can achieve whatever they want, if they really want it. And this coincided with my growing knowledge that **what you think about the most, you attract to yourself**.

Understanding this concept blew my mind when I first learned about it, and it disrupted my progress for several years.

You see, when I was a young teenager, I developed a technique for causing things to happen that I wanted to happen. When I really wanted something to happen, I would think clearly in my mind about the fact that I **betted** that it would **never** happen. This worked so repeatedly well that it became second nature for me. There could be no coincidence that the things that I thought about in this way, more often than not, occurred almost immediately.

For some reason, which I do not fully understand, I simply grew out of it as I got into my later teens. I distinctly remember the feeling that it was childish, and I think I even felt that I should be a little embarrassed by it. This is an example of what I've described before, and I believe, is direct evidence of society gradually taking away and eroding many of our childhood truths from us, as we grow up.

So, when I learned that the **law of attraction** suggested that whatever you thought about, you attracted to yourself, and when I learned further that to maximise the power you should focus with desire on what you **wanted**, I was completely confused by it.

Utterly discombobulated!

How could that work with what I had learned and tangibly quantified in my early teens? Surely it was the other way around?

But then I found the key to my conundrum:

3.2 The Secret

You see, what I learned was that *the law of attraction* **knows no polarity.**

That meant that when I had thought intensively about *not* wanting something, it's exactly the same as being passionate about, and thinking about *wanting* that same something, and that it has exactly the same effect!

Just take a few minutes to think about the significance of that statement.

Does this begin to explain how poor people who think about not wanting poverty all the time stay poor? Does this explain how people with addictions remain in addictions, because they spend so much time thinking about not wanting whatever they are addicted to? Can you see how much easier it would be for someone in abundance to remain in abundance?

It explains a lot of things, including much about human psychology and circumstance. It was a springboard for me, but still it took time for me to really understand what I was later to learn, and to achieve as a result. And I had some very painful lessons that allowed me to see very clearly that *you have to be very careful what you wish for!*

A Theory of Everything

NEAT LIFE HACKS: BE VERY CAREFUL WHAT YOU WISH FOR!

At about this time, I came across the book *Think and Grow Rich* by Napoleon Hill, as advocated by the great Jim Rohn, whom I listened to at the time on a regular basis. This is possibly the most useful and informative book that I've ever had the pleasure of investing in and reading, and it embellished everything that I'd learned previously, and put it all into one concise and comprehensive format that was easy to follow through. For me, rich means wealthy in whatever respect, and not only to do with financial wealth, and success means being your true self.

Book Reference: The Ultimate Jim Rohn Library (tinyurl.com/y98svnl8)
Book Reference: Think and Grow Rich – Napoleon Hill (tinyurl.com/yblckp6y)

Napoleon Hill was commissioned by Andrew Carnegie (American Steel Corporation) in the early 20th century, to perform a study of 500 of the wealthiest business individuals in America, to identify what the difference was between *them* and *'normal'* people.

He identified that the key difference was primarily to do with mentality, attitude, and resulting beliefs, behaviours and actions. Although the book doesn't address it directly, it is absolutely talking about *the law of attraction*, or *Universal Energy*, or *the Secret*, and other globally accepted positive mental attitude techniques and practices. I expect that it didn't mention this explicitly because people back then, and perhaps even now, would have just seen it as so bizarre and outlandish, that they would have switched off immediately, and dismissed the book.

"Truly, 'thoughts are things', and powerful things at that, when they are mixed with definiteness of purpose, persistence, and a burning desire for their translation into riches, or other material objects" Napoleon Hill

"Remember to look up at the stars and not down at your feet. Try to make sense of what you see and wonder about what makes the universe exist. Be curious. And however difficult life may seem, there is always something you can do and succeed at. It matters that you don't just give up." Stephen Hawking

---

What Napoleon Hill teaches is that the starting point of all achievement is **Desire**. Desire is the burning fuel that is the first step towards 'success'. The stronger the desire, the more likely the 'success' will be.

The second step to achievement is **Faith**. Faith means having confidence, trust and an absolute unwavering belief that you can achieve something. He says that if you have a nagging doubt in the back of your mind, or if you are just going through the motions of pretending that you believe, it won't work, because your subconscious will know your doubts.

He refers to the third part of the process as **autosuggestion**. Autosuggestion is a term which applies to all suggestions and all self-administered stimuli that reach the mind through any of the five senses. It is self-suggestion. He teaches that no thought, whether it be negative or positive, can enter the subconscious mind without the aid of the principle of autosuggestion.

What this means is consciously and deliberately orchestrating the scenarios and situations that you engage in, to be environments where you're going to be exposed to the ideas and driving forces that you need to realise your 'success'.

"You are the average of the five people you spend most of your time around" Practical Psychology

---

Part four of the process is referred to as **Specialised Knowledge**. He explains that there are two kinds of knowledge. One is general, the other is specialised. General knowledge, or what we are generally taught at school, no matter how great in quantity or variety it may be, is not of any use in the accumulation of money. Knowledge will not attract money, unless it is specialised, organised, and intelligently directed through practical plans of action, to the definitive end of accumulation of wealth. I do not, however, personally advocate that anyone base their desires on 'money', but focus on the 'thing' of the desire, and the money just happens to facilitate it.

Principle five is **Imagination**, the workshop of the mind! He explains that the imagination is literally the workshop of the mind where we devise and create our plans. Here the impulse and desire come into your vision in the form of required actions. He states that it has been said that man can create anything that he can imagine.

The sixth part is **Organised Planning**. A goal without an organised plan can be viewed as being just a 'dream'.

His seventh principle, and a favourite one of mine, is that of making **Decisions** and avoiding procrastination. And lastly, he suggests that none of what you've achieved in the process during the previous seven will be of much benefit to you at all without **Persistence**.

Equally importantly, he also talks about the influence that the people you hang around with, or have in your life, have on you, and that if you want anything in life, then you should associate yourself with people who already have it. So, if it's money, then hang around with rich people, and if it's happiness and harmony, then find some happy and balanced people to hang around with!

A Theory of Everything

In around 2007 I created my first Mastermind Group. A mastermind group is a peer-to-peer mentoring concept used to help members solve their problems with input and advice from the other group members. The concept was coined in 1925 by author Napoleon Hill in his book *The Law of Success*, and described in more detail in his 1937 book, *Think and Grow Rich*.

It was an interesting experience, and I'm keen to develop another group in the future!

This is a neat and concise overview of *Think and Grow Rich*:

Think and Grow Rich – Napoleon Hill – ANIMATED BOOK REVIEW (youtu.be/FyjLOg8o6_w)

NEAT LIFE HACKS: MONEY IS REPORTED TO BE A MADE-UP CONCEPT USED TO CONTROL THE PEOPLE OF THE EARTH. IF MONEY IS WHAT YOU WANT, AND WHAT YOU FOCUS YOUR DESIRE ON, THEN I'M PRETTY SURE IT WILL ONLY LEAD YOU TO MISERY AND DISAPPOINTMENT. I SEE MONEY AS BEING SIMPLY A TOOL TO GET WHAT YOU WANT (BOB PROCTOR), AND TO SATISFY YOUR DESIRES. SO, IF I FOCUS WITH PASSION AND DESIRE ON WHAT IT IS I ACTUALLY WANT, AND NOT ON THE MONEY, AND IF I HAVE BELIEF AND FAITH IN THE PROCESS (WHICH I ABSOLUTELY DO), THEN THE MONEY JUST HAPPENS, AND FACILITATES WHATEVER IT IS!

"History records that the money chargers have used every form of abuse, intrigue, deceit, and violent means possible to maintain their control over government by controlling money and its issuance" James Madison

---

In my yet unpublished book, 'Who the Fuck, is Jon Tuck?' which is essentially my more in-depth autobiography (because I like writing about myself), I write about a period when I'm transiting from the scientist that I was educated and excited to be, into a more rounded, grounded, and globally educated person I appeared to be becoming, who was more and more interested in connecting the science with spirituality and non-physical phenomena, such as energy.

As I mentioned before, you can't see electricity, but we accept that it's there, because we can see that it makes things work when you switch them on. In the same way, you can't see the wind, or the air that we breathe, but we accept it because we can feel it and see it moving the trees. But what about magnetism, love, energy healing or God, for example. Just because you cannot see it doesn't make it a non-credible phenomenon.

3.3 Unsinkable

Despite my new insights into the law of attraction and thinking I knew how to grow rich; I was also making a lot of big 'mistakes'. Alcohol still ruled my world and it drove a bitter streak between your mother and me. I was going through bankruptcy while trying to start a new business just as you were born, Alex.

These were very confusing times for me, and I will always be deeply and remorsefully sorry to your mother for having dragged her into that mess.

I realise I can't go back and change the past, and I consider all the experiences of that time as essential and valuable learning milestones for me. But all the same, I'm aware that these were the main contributing factors that drove your mother and me apart, to the point that we were happier living apart for your sake, and for us to be able to function as parents.

I'll never forget the day that I was reading a bedtime story to you, William, and I became overwhelmed with emotion and had to stop reading and leave the room, to prevent you from seeing me cry (which I've since overcome, and you see my face leaking sometimes).

A Theory of Everything

I was at the lowest that I'd been for a long time, and quite desperate. I was confused with all the information I'd been absorbing, the bankruptcy and lack of career and life focus, and the breakdown in the relationship with your mother, whom I'd loved so dearly. The lack of clarity was taking its toll. At this time, I thought I was doing everything I could to progress my career, education, relationship, and self-development, but nothing seemed to be working. I was at the end of my tether. Little did I know it, but right there and then, I was exactly '**three feet from gold**'!

NEAT LIFE HACKS: WHAT I HAVE FOUND CONSISTENTLY IN MY LIFE IS THAT WHEN YOU REALLY WANT TO ACHIEVE, OR DO, OR GET SOMETHING, AND YOU FOCUS YOUR DESIRES ON IT, WITH BELIEF AND FAITH, FEELING AND VISUALISATION, LIFE WILL OFTEN THROW A WHOLE TON OF CONFUSION AND CHALLENGES AT YOU, JUST BEFORE THE POINT OF BREAKTHROUGH AND ACHIEVEMENT. HENCE THE EXPRESSION '**THREE FEET FROM GOLD**' (TAKEN FROM *THINK AND GROW RICH*). WHAT I'VE FOUND IS, IT'S LIKE LIFE IS SAYING, 'ARE YOU *SURE* YOU REALLY WANT THIS?' AND IF YOU REMAIN STRONG AND FOCUSSED, AND PLOUGH THROUGH THE CHALLENGES, THEN IT WILL SEE THAT YOU ARE SERIOUS, AND IT WILL PROVIDE IN ABUNDANCE! ALSO, 'THREE FEET FROM GOLD' IS A GOOD 'WHAT IF, UP' STATEMENT (SEE 6.7)!

It's a bit like when, William, you had the dream and desire to reach the town football A-team, after you'd progressed so well in the town B-team. Your opportunity presented itself with a scheduled match against the A-team, but it was a terrible game where you were played out of position and were beaten 5:1. You were gutted and emotional, but you stayed strong, and right there and then you were three feet from gold! Within a week you got the call to join the A-team, and during your first game with them, you not only beat your main rivals in the league, but you contributed nicely to hammering them 7:1!

So, when I left your room during the bedtime story back then, I went and sat at the computer in my room and I looked up towards where the sky would be if there wasn't a roof in the way, and I asked for help. Like I was praying. I asked for a sign. I asked for support. I asked, 'What should I do now?' I was desperate.

Through my tears, I noticed that an email just popped into my inbox, and I couldn't believe it when it said, 'Are you sitting in your room, looking towards where the sky would be if there wasn't a roof in the way, asking for help?'!

Well, it didn't say exactly that, but as good as!

Now, I'm already acutely sceptical at this stage of so-called 'get rich quick' and 'make your willy as big as Nelson's Column' promises on the internet (which were even sent to your mother), and I approached this with caution, too. But at the same time, **I don't believe in coincidences**, and wanted to find out as much as I could first, before I dismissed it.

NEAT LIFE HACKS: THERE IS NO SUCH THING AS COINCIDENCE. EVERYTHING HAPPENS FOR A REASON, AND MORE OFTEN THAN NOT, BECAUSE YOU, OR SOMEONE ELSE HAS ATTRACTED IT INTO BEING OR HAPPENING. FOR THE SAME REASON, I DON'T BELIEVE IN 'LUCK' AS WE ARE LED TO UNDERSTAND IT. IT IS MY OPINION THAT FOR THE MAJORITY OF CASES, YOU CREATE YOUR OWN 'LUCK', AND I OFTEN WISH PEOPLE THE BEST OF 'TUCK'!

The email advocated a webinar called 'Unsinkable' by Sonia Ricotti, and the webinar was at some arbitrary time in the US; but when I checked, it was in five minutes' time, local time! Well, it was free, I had nothing to lose, and so I logged on and tuned in to find out more.

This turned out to be one of the best 'risks' I've ever bought into in my entire life, and what I was to learn, and also invest in, was one of the most valuable products and experiences that I'd come across during this period for me. The timing was simply perfect! It's not going to be for everyone, I accept that, but it was perfect for me! And *I don't believe in 'coincidences'*!

This was to kick-start my understanding and practice of meditation, which has now become an integrated part of my life for me. It was to kick-start my understanding of stress and anxiety and to build upon my ability to remove my negative thoughts and emotions. It was to kick-start my understanding of accessing and influencing my deep subconscious. It was to kick-start my appreciation of synchronicity. It was to kick-start my fascination for energy healing, and training in Reiki. And it was to kick-start the gradual rebuilding of me!

"Meditation makes you innocent, it makes you childlike. In that state, miracles are possible. That state is pure magic. A great transformation happens – In innocence you transcend the mind, and to transcend the mind is to become the awakened one, the enlightened one" Osho

"True healing will always begin with your thoughts. Master your thoughts and you will master your life" April Peedess

"If every 8-year-old in the world is taught meditation, we will eliminate violence from the world within one generation" Dalai Lama

---

So, what can I briefly teach you of what this remarkable programme taught me, and what I practise routinely to this day, as a result of what I learned, and what it inspired me to subsequently research, pursue, and master?

3.4 Worrying About Stuff

First it taught me that there's no point worrying about stuff you cannot change:

"Please grant me the serenity to accept the things I cannot change, the courage to change the things I can, and the wisdom to know the difference" The Serenity Prayer

"This, too, shall pass" Sonia Ricotti

'This, too, shall pass' – Sonia Ricotti! And it's true, and it makes so much sense. What's the point of 'worrying' about anything, unless the focus is changed to a constructive approach towards what you're going to do about it?

"Worrying merely impairs your ability to maximise the impact you have on situations at hand – chill out!" Mr. Jonathan R. Tuck (aged 16)

A good rule of thumb that I've come across is to consider whether 'it', whatever the challenge is, is going to matter in five years' time, and if not, then it's not worth investing any time or energy into thinking too hard about it.

3.5 Forgiveness, Gratitude, Joy and Unconditional Love

Secondly it taught me the significant and remarkable value of *forgiveness, gratitude, joy and unconditional love:*

3.5.1 FORGIVENESS

This is a very powerful tool. I realised that when you hold a grudge against someone or something, *you are only hurting yourself*. And there's no point in that. What I found was that the more I forgave the people that I saw as having hurt me in the past (including myself), I became freer and happier in general, and more able to move on, and make a better future for myself.

This was also directly in harmony with my belief (as a result of my studying human psychology) that most people in general are not out to hurt anyone else, and that there is always a very good fundamental underlying reason as to why they behave the way they do, and it's not necessarily their fault or intention.

I don't blame anyone for their behaviour. I hold them *responsible*. That's a very different sentiment.

This is an exceptionally useful tool, and not just in forgiveness (please see the link below):

Ho' oponopono:

I'm sorry
Please forgive me
Thank you
I love you
Understanding the ancient Hawaiian practice of Forgiveness (tinyurl. com/y6vzckfr)

3.5.2 GRATITUDE

Now this is a big one that should never be overlooked! Quite simply, the more grateful you are, and I mean sincerely heartfelt gratitude, then the more good you attract to yourself. I struggled with this for a while, but I knew that the gratitude had to be sincere, and not just as part of a tool to attract more abundance to myself. I kept a gratitude manifesto, which I read and added to every morning, before it gradually became second nature to me to naturally observe my gratitude for everything around me, every day, and for everything I have or experience.

The (quantum) physics behind it suggests that as you feel and express gratitude, you raise your vibrational energy. In doing so, you put yourself in a position of receiving energy at that higher level. The higher the vibrational energy, the more positive and fruitful it becomes.

"As you begin to be grateful for what others take for granted, that vibration of gratitude makes you more receptive to good in your life" Michael Bernard Beckwith

"It is through gratitude for the present moment that the spiritual dimension of life opens up" Eckhart Tolle

"Gratitude, like faith, is a muscle. The more you use it, the stronger it becomes, and the more power you have to use it on your behalf. If you do not practice gratefulness, its benefaction will go unnoticed, and your capacity to draw on its gifts will be diminished. To be grateful is to find blessings in everything. This is the most powerful attitude to adopt, for there are blessings in everything" Alan Cohen

---

3.5.3 JOY

Joy is a very powerful and uplifting state to be in, and to endow upon others. It will raise not only your own vibration, but also the vibration of those you share it with.

"You know, the ancient Egyptians had a beautiful belief about death. When their souls got to the entrance to Heaven, the guards asked two questions. Their answers determined whether they were able to enter or not. 'Have you found joy in your life?' 'Has your life brought joy to others?'" Morgan Freeman

---

3.5.4 UNCONDITIONAL LOVE

The word *'unconditional'* is of paramount importance in this statement.

I think that to love someone or something unconditionally is one of the most beautiful states a human being can find themselves in. In our society, there are a plethora of challenges surrounding this, most of which appear to have been orchestrated deliberately so as to erode and destroy the significant power of love.

"All you need is love" John Lennon

In my opinion, **love is the most powerful energy in the Universe**, and if us plebs all got together and focussed love on evil, I'm absolutely convinced that love would kick evil's arse! This is akin for example to the Jedi against the Empire in *Star Wars*. 'The Force' for me represents the energy of 'love'. Feel the force, Luke…

Interestingly, I was one of the many who became a 'Jedi Knight' after the 2001 UK Census ruled 'Jedi' as being an official religion, and I'd now consider myself to be a Jedi Master (Reiki Master).

"I believe in the kind of love that doesn't demand me to prove my worth and sit in anxiety. I crave a natural connection, where my soul is able to recognise a feeling of home in another. Something free-flowing, something simple. Something that allows me to be me without question" Joey Palermo

"The Universe only pretends to be made of matter. Secretly, it is made of love" Daniel Pinchbeck

---

Love is apparently a measurable and tangible energy, which vibrates at a frequency of 528 Hz.

3.6 Don't Take Anything for Granted

To embellish the importance of gratitude further, we seem to mostly just **take things for granted**. I believe that in the Western world and civilisation, it is very easy to overlook just how grateful we could be, and should be, for the things we have, the life we're able to live, and the opportunities that we have, if we want them.

So, despite my views on the *'system'*, actually I'm acutely aware that what we have in our civilisation and society, through whatever means, is almost completely unique compared to the standard of living of many other people around the world. The things we take for granted at a fundamental level are the fact that we have fresh water to drink, we have warm beds at night, and we have homes that we can call our own. And more importantly, *we are safe*. Relatively speaking of course. And all this, in spite of the proliferation of fear and hatred based on the divide and conquer principle that we are subject to through the media and other sources.

The moment that it became crystal-clear for me just how grateful I actually was to be alive, in the body that I possessed, and in the environment that I was alive in, and to have had the opportunities that I had, and to be as privileged as I was, was a very special moment indeed, and exceptionally profound.

NEAT LIFE HACKS: DON'T EVER TAKE ANYTHING OR ANYONE FOR GRANTED. NOW THAT'S CHALLENGING, BECAUSE WE LIVE IN A SOCIETY THAT HAS TAUGHT US TO TAKE PRETTY MUCH EVERYTHING FOR GRANTED. EQUALLY, DON'T RELY ON YOUR EXPECTATIONS. IN FACT, THE BEST POSITION IS: DO NOT HAVE ANY EXPECTATIONS, BUT BE AWARE OF OPPORTUNITY AND POSSIBILITY NEVERTHELESS. BE ADAPTIVE AND INNOVATIVE TO CHANGE, BECAUSE THAT'S THE ONLY CERTAIN THING IN LIFE (OTHER THAN DEATH) – CHANGE IS INEVITABLE, AND THOSE WHO RESIST IT WILL SUFFER IN THE LONG TERM

NEAT LIFE HACKS: I FIND GREAT PLEASURE IN DOING THINGS FOR OTHERS AND HELPING OTHERS. SOMETHING THAT I LEARNED IS THAT WHAT GOES AROUND, COMES AROUND. WHAT I FOUND IS *CRITICALLY* IMPORTANT TO UNDERSTAND ABOUT THIS STATEMENT IS THAT GIVING MUST BE UNCONDITIONAL, AND ONE MUST NEVER EXPECT TO BE GIVEN BACK DIRECTLY AS A RESULT, OR TO EXPECT ANYTHING IN RETURN DIRECTLY FROM THOSE THAT YOU GIVE TO. YOU SEE, THE WAY IT WORKS IS THAT IF YOU GIVE (OR PAY IT FORWARD), YOU WILL GET YOUR KINDNESS OR ENERGY OR LOVE BACK, BUT IT MAY COME FROM A DIFFERENT PERSON, ANGLE OR SOURCE

3.7 Regular Morning and Evening Meditation

The Unsinkable package also provided me with short, 15-minute guided audio meditations for the morning and the evening. I found these to be invaluable. The morning meditation uses your intuition to set you up for the day, and to establish what it is you want, or need to achieve, or focus on for the day, and the evening meditation removes all anxieties experienced during the day and assists you with a good night's sleep.

3.8 Binaural Beats-Assisted Meditation

'Unsinkable' started my journey in to meditation, and led succinctly to my discovery of audio-beats or Binaural Beats. When you're listening to something recorded with binaural beats, you hear different sound frequencies in the right and left ear. It's postulated that this has the effect of bringing you quite quickly into a meditative state and can be utilised to directly access and influence the beliefs stored in your deep subconscious. This is exactly what I was looking to find!

MY LAST STATEMENT IS A DELIBERATE OXYMORON! I HAVE LEARNED THAT THE LANGUAGE THAT YOU USE IN YOUR OWN THOUGHTS AND COMMUNICATION WITH OTHERS IS VASTLY POWERFUL. SO, WHAT I MEAN HERE IS THAT I HAVE MIXED THE USE FOR THE WORD 'LOOKING' AND 'FINDING', AND IN TERMS OF USING THE LAW OF ATTRACTION, THERE IS A BIG DIFFERENCE BETWEEN THE TWO. IF YOU SEND OUT THE MESSAGE TO THE UNIVERSE THAT YOU ARE 'SEARCHING', OR 'LOOKING' FOR SOMETHING, THEN IT WILL PROVIDE YOU WITH EXACTLY THAT, A SEARCH! WHEREAS I HAVE FOUND THAT IF YOU PRESENT THE CASE THAT YOU ARE IN THE PROCESS OF 'FINDING' SOMETHING, THEN IT WILL PROVIDE IT TO YOU FAR MORE SUCCINCTLY. THIS WAS A FUNDAMENTAL EXPLANATION THAT WAS GIVEN TO ME BY CHRIS NEW EARLY ON IN MY QUEST FOR UNDERSTANDING.

A Theory of Everything

The programme that I invested in, as advocated by Sonia Ricotti, used Binaural Beats, and promised *'Awakening'* and reprogramming of the subconscious. It was a lengthy process, and it saw me meditating for an hour every evening, which got on your mother's nerves a bit. But I persisted, and in the early phases of about a year, I saw some measurable results in the way I saw life, my behaviour, and the way I responded to my environment. I wasn't drinking at this time either, which helped, if not enabled me to engage the practice, and gain full advantage!

Reference: CenterPointe Holosync

Sadly, I stopped the programme at around level two, the reprogramming, where I'd recorded my own affirmations which were subsequently subliminally superimposed over the tracks. I stopped after about six months of this, having seen some good results, and it's no coincidence that this shift of attention coincided with me not making time as a result of the take-off of my next career, and the realisation and materialisation of my wildest dreams: my adventure as co-founder and chief scientific officer of a company which develops and produces ultra-fast charge energy storage devices, or supercapacitors.

One of the things I learned from all the research and subsequent experience that I was to enjoy as a result was that if you really want something, then you have to be specific – and again, for me, it's not all about the money. In fact, I'm finding that I'm not any longer very money motivated, or certainly not as much as I was in my early career.

"Money is merely a tool to facilitate the things of your desires"
Bob Proctor

Another thing I've learned is that you must be very careful what you wish for!

"Be very careful what you wish for" Dr Jonathan R. Tuck

One of the most concise and useful resources I've come across, which explains the law of attraction and how it works from a scientific perspective, and provides a very structured and effective way of identifying and documenting, and attracting your desires, is given in Michael Losier's book:

Book Reference: Law of Attraction: "The Science of Attracting More of What You Want and Less of What You Don't" by Michael Losier (tinyurl. com/y7odjknm)

Michael not only uses the original principles of 'contrast' first presented by Bill and Ester Hicks to define what you really want, but also presents very clear strategies to mitigate your subconscious blockages that would otherwise prevent the law of attraction from working. I have personally used this method with some very tangible results, in the absence of subconsciously attracting anything of what I don't want at the same time!

Part 4
The Rebuilding of Me

4.1 Patience and Persistence

Shortly after my breakdown over the bedtime story that evening, and embarking upon the 'Unsinkable' programme, in conjunction with what I was learning in the book *Think and Grow Rich*, and had learned from *The Field*, and *The Biology of Belief*, and using all of the information I was gradually gathering, I put together a wish list for my next career, and I sent it out to the Universe, in the same way you might be encouraged to send out your Christmas list to Santa.

It was specific (in the same way your Christmas lists are always very specific!). It covered exactly what I thought I wanted to do, where I thought I wanted to do it, in what field I thought I wanted to do it, with the type of people that I thought I wanted to do it with, what remuneration I thought I wanted to receive for doing it, and when I thought I wanted it to happen.

I was not afraid to be bold, or what I felt was being bold at that time. I was not afraid to reach for the stars. I was not afraid to challenge myself to step up to all that I knew deep down I was capable of, but just hadn't been given the opportunity to prove yet.

A Theory of Everything

NEAT LIFE HACK: DON'T BE AFRAID TO WISH FOR WHAT YOU REALLY WANT, AND NOT JUST WHAT SOCIETY LEADS YOU TO PERCEIVE AS BEING SOCIALLY 'REASONABLE' OR 'REALISTIC'

It 'happened' almost immediately. By this I mean that, within a week, my business partner-to-be approached me to see if I was interested in working with him to start what was soon to be our new and very buoyant company.

The only element of this process that was unclear to me, or that didn't pan out as I expected, was the 'time' element, or the date I associated with my expectation of when it should happen. I think now that I understand it, this is possibly explained in a number of ways.

For example, firstly, it takes 'time' for the Universe to line everything up properly. Especially when it's likely to be receiving mixed messages from you as a result of your underlying confused subconscious beliefs, thoughts and emotions. **Patience and persistence** are key.

Wow, patience was a big one for me! I was inherently, and for whatever reason still slightly am today, very impatient. That took a lot to master!

"When I run after what I think I want, my days are a furnace of stress and anxiety; if I sit in my own place of patience, what I need flows to me, and without pain. From this I understand that what I want also wants me, is looking for me and attracting me. There is a great secret here for anyone who can grasp it" Rumi

---

Yeah, Rumi, that's all very well and good, but I think, as Jim Carrey so clearly notes in his explanation of the story of his success, you can't just wish for something and then just sit on your arse (or, 'fanny' in American) and wait for it to happen. You must react to the messages and opportunities that the Universe is providing, and take appropriate action, sometimes in small steps, in order to move forwards. I think I understand what you're saying, though!

Secondly, 'time' actually appears to be a made-up concept that was invented with the primary objective of controlling us plebs, and organising or structuring the maximum amount of work that 'they' can get from us. It's interesting to note how differently other civilisations and cultures are reported to have viewed the concept of time over the millennia, and how their days and active hours are structured very differently. According to nature's rhythms as opposed to alarm clocks!

For example, back in the 18th century, it was commonplace for people to have a sleep during the day and then stay up late into the night and early morning. Rather like the modern-day siesta that is common in hot countries. I know from my own experience that these late-night hours can be some of the most productive ones of the day, and some meditations are specifically advocated as being more powerful and enlightening when performed in the early hours of the morning (around 04:00).

My overlying understanding surrounding the concept of time is that:

---

*"There only **is**; has only ever **been**; and only ever will **be**, **right here and now"*** Dr Jonathan R. Tuck

---

By this, I mean that: the only time you will **ever** experience is **right *here and now***.

The past is in the past, and the future (as you perceive it) hasn't happened yet.

It's not worth beating yourself up about what's happened in the **past**, because you cannot change it. You're best positioned to look at it as a positive learning experience from which you can apply what you learned to the now.

It's not worth worrying about the **future either**, other than envisioning the positive experience that you're going to gain in the *future*, by applying the results of the knowledge that you acquired from the *past*. In fact, you are actually creating your future, as a reflection of who you are in the now.

It's interesting to contrast this understanding with that of the Incas, for example. They saw time in three dimensions, all running simultaneously with one another, past, present and future. Something which is emanated in the teachings of Dolores Cannon, and widely adopted in Eastern culture. This perspective also begins to explain how for example in Reiki, the practitioner can perform past time-line healing, and past life healing, and can influence your future needs.

Please also see a very concise and well-presented overview by Jesse Elder:

Reference: Time is a Useful Illusion (youtu.be/DR5aYgcch8Q)

4.2 Do It Now!

So, my philosophy on utilising the essence of right here and now, is, if you are going to do something, then, within good reason, **do it now**! Don't wait! *Procrastination is a terrible burden*, and leads to a downward spiral – see Parkinson's Law

Book Reference: 'Screw It, Let's Do It' Richard Branson (tinyurl.com/y72sncea)

One recent example of this principle that I have, is during our experience of sailing. From the very beginning, I openly accept that, as much as I wanted you to see if you enjoyed it or not, it was *my* dream, not yours.

So I wanted to try to get you into sailing. I had a little Mirror Dinghy at a club in the Cotswolds, and wouldn't it be great if we could enjoy a family day sailing?

A Theory of Everything

When I first took you to the club and we tried to sail it, I immediately learned two things:

Firstly, that you were not impressed in the slightest with my helmsmanship. I got all tangled up in the sheets and we set off quite unpredictably in the wrong direction. This, on top of the previous two times I'd taken you out in a powerboat and subsequently had to be towed back to shore by the safety boat, both times because I broke the propeller, didn't really support your trusting me as your captain on the water.

Secondly, I realised that, even though I had lost lots of weight and felt more physically mobile than I had done for a long time, I was still not as mobile as I was when I was in my early twenties when I sailed and crewed on a regular basis. I had to accept the fact that as a result of my deteriorating mobility at that time, I could no longer sail safely, and that I certainly was not going to put you boys in any unnecessary danger.

William, you seemed to take to sailing very well, and appeared to be enjoying it. But Alex, you wouldn't be told what to do, and belligerently sat in the car all day with people walking around the club talking about you as if you were a caged puppy. That was a challenging one for me!

But when we took the larger sailing boats and powerboats out, you joined in and seemed to have great fun in the process.

So, after a discussion with one of the safety officers I realised that I had the opportunity to address my apparent incompetence in powerboating if I became a safety officer. I could then entertain, you, too, Alex, whilst William was sailing. **There and then I formulated a plan**!

I screwed it, and I did it now!

I booked my Powerboat level II course immediately, and within three weeks I had passed, and had completed the club safety officer course. This meant that during the very next session, I was able to take you out on the water as an Assistant Safety Officer.

Well, as you know, it didn't pan out exactly as I'd hoped or expected, because William, within a month you quit sailing in favour of your true passion that was football, but that's all OK. It at least gave you the opportunity to see if you'd enjoy something that I'm interested in, and, as a result, I now have the ability and licence to daytime charter yachts and powerboats around world, if the desire takes me.

My point is, though, that I didn't procrastinate or think too much about it. I saw the opportunity and I just got straight on with it. And I think, equally importantly, that I have no regrets and am glad that I did it, and I know that I wouldn't have considered it, or had that opportunity, if it were not for Alex sitting in the car all day, listening to music and adjusting all the settings on the dashboard!

Equally, William, when you turned around and told me that you no longer wanted to go sailing or be a part of the club, for whatever reason, I didn't get angry or animated, I simply turned my attention to how grateful I was for the experience that it had provided us both with, and I moved on quickly. As you're aware, I'm currently in the process of applying this principle to other situations where I might lose my rag, for whatever reason.

4.3 Making Decisions

The last story leads me to what I learned about making decisions. And I can only tell what works for me, and why I believe it to be the most appropriate advice. Some people may disagree.

My philosophy is:

---

*"Make decisions **quickly**, and change them **slowly**."* Bob Proctor

---

I first learned about this principle from Bob Proctor, and I think that it's an exceptionally important principle that relies heavily on your immediate

Intuition. Many people are unaware of the significance of their intuition, or simply do not trust it, for whatever reason.

This statement is also based on the principle that it is better to make a bad decision than not to make a decision at all, as we mostly learn from our mistakes. It also ties in nicely with my views on procrastination, and the importance of *doing it now*!

NEAT LIFE HACKS: WHEN CONFRONTED WITH A SITUATION WHERE YOU HAVE TO MAKE A DECISION, HAVE FAITH IN YOUR INTUITION AND GO WITH THE FIRST THING THAT SPRINGS TO MIND, OR THE THING YOU FEEL MOST STRONGLY TOWARDS, EVEN IF IT SEEMS COUNTER-INTUITIVE. MORE OFTEN THAN NOT YOU WILL HAVE MADE THE RIGHT DECISION; HOWEVER, IF YOU SUBSEQUENTLY FEEL AS THOUGH YOU COULD HAVE MADE A BETTER DECISION, BE GRATEFUL FOR THE LESSON THAT YOU LEARNED, AND APPLY THE GAINED KNOWLEDGE IN THE FUTURE

4.4 IVVM

Having only used the process of Law of Attraction a few times in the past, with varying degrees of success, I wasn't convinced that as a result of sending my request to the Universe, the connection I had made with this new potential business partner was the one. So I launched Recruitment Campaign 3.2 (career 3, part 2) and spent a good year and a half belligerently scraping the recruitment scene, and working in various interim positions, including at the same company as your mother, whilst working evenings and weekends with my business partner (which got on your mother's nerves some more).

Interestingly, at this time, I also spent lots of time working with your Scout Group, William, as well as with the South Wales and Western Polymer Group, to try as best as I could to keep as close to my field of 'expertise' as possible.

A Theory of Everything

Every morning I read the document that I had produced, and described earlier, and practised what I had learned to be **IVVM**.

IVVM is a quick and simple affirmation technique, which is practised in various forms by many of the greatest (admirable) and successful athletes and people on the planet. It's like PMA, or Positive Mental Attitude – If you want to succeed, then think positively about what you want to achieve, and your ability to achieve it, visualise it, open yourself up to the possibility, and then draw the eventuality to yourself.

NEAT LIFE HACKS: IF YOU'RE PRACTISING THE LAW OF ATTRACTION AND AFFIRMATION, THEN THERE ARE A NUMBER OF KEY INGREDIENTS THAT YOU MUST EMPLOY, AND ONE OF THOSE IS **ACTION**! YOU CAN'T DO AN AFFIRMATION AND ASK FOR SOMETHING, AND THEN SIT ON YOUR ARSE DOING NOTHING WAITING FOR IT TO HAPPEN! IT'S NOT GOING TO HAPPEN THAT WAY! YOU MUST TAKE SOME ACTIONS AND STEPS TOWARDS YOUR GOALS, NO MATTER WHAT THEY ARE

So, **IVVM** stands for…

Idea, or the idea of what you want to do, be, or achieve;
Visualisation, or visualise what it is you want to achieve;
Verbalise, or see yourself verbalising what you want to achieve, and;
Materialisation, or see *and* 'feel' in your mind the materialisation of what it is that you want to achieve

In this process, in the visualisation, size, movement and colour are all important. Of exceptional importance is to experience **the feeling** that you would have if you were already in the position of having it, or having achieved it. This is quite critical, and in line with one of the most profound things that Diado (Mum's dad) told me very early on, and that was that if you wanted something, then you should act and feel as if it has already

109

happened. From what I understood, he had also read a variation of *Think and Grow Rich* and we shared several core principles and ideologies, and he had certainly made his mark on the business success front, too!

Another important aspect of this technique is right at the end during the materialisation. During the materialisation, the result is more profound if you imagine the result being drawn towards you (Jill Sudbury) and truly experience how happy it makes you feel.

All this made a lot of sense to me, and even the time element of it made some sense. You see, I've always believed that if you really want something, then as long as you're taking small steps towards it, then, as tortuous as the path may be, you'll always get there eventually, with persistence, dedication, action, patience and 'time'.

For over a year I IVVM'd my arse (or fanny) off every morning. I didn't want to take anything for granted, and to this day, I'm not sure if having sent the original message to the Universe was actually enough, and I didn't need to do any more, or whether the continual banging on to the Universe about what I wanted had any effect at all. This is the first time that I had practised it on something as important or as big as this, and I wasn't sure exactly what the rules were. From past experience, I knew that the path didn't matter, it was the intention, vision, belief and faith in the process. The feeling of it having already happened, and the persistent action that counted for everything.

4.5 Eat a Frog for Breakfast

One of the significantly important things that I learned during this time as well was to 'eat a frog for breakfast'!

Now, of course, this is a metaphor, and doesn't actually mean that you should serve up some frog on toast for breakfast, with a siding of lizard tongue and salamander gizzard!

What it's recommending is that you should look at your daily tasks in the morning and choose the most ugly, unpalatable and challenging task, and do that first!

I've found this to be a very useful principle on multiple fronts, because it gets the task that you're most likely to procrastinate done and dusted, and out of the way early on, when you're more likely to be able to focus better, and, as importantly, it gives you a very good sense of achievement early on in the day. Like making your bed in the morning!

4.6 Emotional Freedom Techniques (EFT)

As part of the rebuilding of me, I wanted to overcome my inherent fear of public speaking.

I love the sound of my own voice! I love talking to people about *me*. And I love being able to try to help people with some of the knowledge and experience that I've gained. On any topic, be it scientific or otherwise. So why, oh why was I so apparently terrified, no... *petrified*, of public speaking?

I battled with this for the majority of my adult and professional life. In my late teens, I was acutely aware that it was something that I just had to overcome. It was so important for my career and life ambitions. From my early career, I regularly challenged myself to speak in front of an audience, entering various lecture competitions, and eventually winning the regional heats during my first degree.

I'd presented to audiences greater than a hundred at various arenas around the world, and on numerous occasions, but for a long time it didn't get any easier, or any more comfortable. I found that if I went to a conference, I couldn't concentrate, absorb or enjoy anything prior to giving my talk, which was frustrating, especially since once I got onstage, I usually did a great job.

One of the advances I made in the medium stages of this process was with tapping, or Emotional Freedom Techniques (EFT). Here I worked with a friend of mine, Paula Reed, for several months. Together we traced my blueprint fear, stored in my deep subconscious, to a memory of an event where I was taking the reading at your great-grandfather's church in Bontddu, at the age of about eight or nine. I remembered that I stumbled on the words and was so embarrassed that I ran away and hid in the bell tower.

EFT appeared to have a positive effect on my confidence, and provided me with the foundations of what I was to build upon in the future, including my understanding of how to go back and heal the past, based on the principle that it is actually happening in tandem with the present and the future.

To explain briefly, EFT is a technique which supports us in changing unhelpful subconscious beliefs by using a specific set of meridians in the body and tapping these while stating the unhelpful beliefs. It sounds too simple to be true, but it really does work and there are plentiful case studies to show that it does.

A good resource for case studies and to get more information on EFT is: EFT Universe (eftuniverse.com)

4.7 Three Deep Breaths

I'll never forget one piece of advice on public speaking that I belligerently overlooked for all this time, which was given to me by my professor during my first degree, Professor Page, and that was the utilisation of **three deep breaths**.

'Three deep breaths' is a well-practised and documented method for significantly calming your nerves. The *'Deadliest Catch'* guys even use it. And doing the jobs they do, whatever techniques they use to prevent themselves from dying, I take very seriously! Like knocking on wood!

Fortunately, I was at a stage in my life where I saw the futility in beating myself up for anything, but I'm sure that if I'd have realised sooner that this technique would almost be 100% effective, I'd have physically dislocated one of my legs and properly kicked myself in the butt with it!

So, now I have several warm-up techniques which allow me to overcome my fear of public speaking. This may just be necessary for me. You may not have such a big issue with it, but this is just one example of how, by consciously changing the way you think about something, it can have such a profound effect on how you feel about it, and your subsequent behaviour:

- I change the way I think about the event prior to it, by **consciously thinking positively about how much I'm looking forward to it**, and how I will enjoy every minute of it (whereas before I would think about it with dread, fear and negativity)
- I set my intention to deliver a well-received talk that I'm going to enjoy giving and I **visualise** myself on the stage, and **hear** myself speak on the stage, in a positive light, and actually **feel** myself having that great experience
- Shortly before going onstage, I remove all negative emotions, thoughts and feelings, using my developed techniques, which I will describe in more detail later on
- I take **three deep breaths** immediately prior to going onstage, or immediately before I begin my talk, and actually **physically feel myself relax**
- I interact immediately with the audience, perhaps with a joke, or ask them a question early on in the talk, and maintain audience participation throughout, if appropriate

A Theory of Everything

This is probably a little overkill, but frankly, I'd rather overdo my approach than underdo it!

Equally, your mother has recently completely overcome her fear of public speaking, from what I understand, by simply telling herself that *what she has to say is important.*

NEAT LIFE HACK: IF YOU EVER FIND YOURSELF IN A STATE OF FIGHT-OR-FLIGHT, AND YOU WANT TO LEVERAGE THE FIGHT, RATHER THAN THE FLIGHT, THEN SIMPLY TAKE THREE DEEP BREATHS BEFORE YOU ENGAGE, AND THINK POSITIVE THOUGHTS ABOUT HOW YOU SEE YOURSELF ACHIEVING WHAT YOU WANT TO DO. FEEL HOW HAPPY IT MAKES YOU

Deep breathing is generally a very useful tool to quickly get you into a relaxed state at the beginning of a meditation session, before you relax into the harmonious shallow breathing. It essentially enables your body to take in more oxygen in order to help all areas of your body work better, including your brain and mind. This is one of the many reasons why people who do sport or exercise tend to be generally fitter and healthier, and they tend to feel better than people who do not do any exercise, because they are naturally encouraged to breath more efficiently as a result of doing the activity.

NEAT LIFE HACK: I FIND IT EXTRAORDINARY THAT ONE OF THE MOST FUNDAMENTAL MECHANISMS OF LIVING: *BREATHING*, IS TAKEN FOR GRANTED AND ALMOST IGNORED AND MISUNDERSTOOD BY MANY OF THE POPULATION, WHO SEEM TO BE PRONE TO SHALLOW BREATHING FOR THE MAJORITY OF THE TIME. BREATHE DEEPLY WHENEVER YOU CAN, AND ALLOW YOUR BODY TO ABSORB THE OXYGEN IT RECEIVES

One very profound meditation that I've done with my group at the Wellbeing Clinic, however, is based on rapid breathing, and is the method used by free divers to expand their lungs and hold their breaths for such a long period of time. Apparently, the world record is held by a guy called Tom Sietas, who held his breath during a dive for 22 minutes and 22 seconds!

One example is an established practice is called the Wim Hof breathing method, and it's incredible! Please see a link below:

Reference: WIM HOF BREATHING TECHNIQUE – How To Breathe Like The Iceman (youtu.be/bLj9s7xQk4I)

Another technique commonly used in meditation is to slow your breathing down, taking long breaths in and slow breaths out. This triggers a signal in the brain that enables you to enter a deeper state of meditation more quickly.

4.8 Materialisation of a Dream

So, going back to my IVVMing and use of the Law of Attraction or Universal Energy, and despite the chaos of a start-up business, by mid-2015 I could now tick every single box on my four-page wish list in the career that I had apparently orchestrated for myself, and thus proved to myself conclusively that this was indeed a very real phenomenon, however it had worked, or whatever I'd done.

And for sure, I could confidently say that I'd had one of the most exciting and fruitful career adventures of my life to date! More importantly, I didn't see it as 'work'. I was playing at what I wanted to do, and I had tremendous fun at it. I've written briefly about my career story in a separate text (published on Linkedin), if you're interested:

Reference: A Start-up Roller Coaster Adventure in Energy Storage (tinyurl. com/ya55ryjt)

Part 5
Relationship Roller Coaster Ride

5.1 Kate and My Road to Recovery from Alcoholism

By mid-2016 I'd grown more and more confident in utilising the Law of Attraction and all my other life-shaping tools. So I started thinking, 'Right, if could do this for my career, I wonder if I can do this for a new relationship?'

Your mother and I had been separated for over a year, and after enjoying my bachelor period a little too much, I felt that I was ready to make a commitment again. I wanted to be loved again, and more than anything, I wanted to love again. To share my life with someone I cared about.

When I met Kate, I was not drinking. To be fair, I was on the cusp. I was just about approaching my target weight, and it was nice to have a glass of wine or a cider with her. And to begin with, that was all it was. Kate moved in with me, with her son, in July 2016, and it was a beautiful summer, in a beautiful part of the world. For the first two weeks at least, I felt that I had made it! We had a lovely 5-bedroom house in Poshville, England, and great neighbours, and Kate's son had got into a great school! We were a little closer geographically to you boys, and we had you with us for the entire duration of the end of the summer holidays that year. I also had a great job locally, with a very large salary to support me!

And then 'bang!' overnight I was made redundant.

It wasn't as bad as it sounds. When starting out, I'd written a year's notice into my contract, and that's exactly what I got, and I'm grateful for that.

However, it also meant that I was now pretty much on paid holiday for a year. The drink flowed fluidly from then on, and the occasional almost terminal arguments ensued. I understood that our arguments occurred as a result of a number of factors, and drink was certainly one of them, but I still didn't want to believe that stopping drinking would solve the problem. I'd already established years ago the fact that if I took a lot of Gaviscon prior to drinking, I didn't get so drunk, and the ramifications were not so bad. So, over the course of six months I worked very hard to try to find a way that I could drink, and for it not to compromise our relationship.

We established that I was not good with red wine. I was gutted by this as I loved red wine, but if that's what it took to keep drinking, then I could deal with it, and from then on drank white wine, which I didn't like very much, and Polish nutty vodka. And then I discovered a wine filter that not only aerated the wine, but also removed the sulphur dioxide preservative that is used in wine to keep it fresh; and when I tried it with my favourite red wine I was able to convince myself that, actually, I could now safely drink red wine without rendering myself an utter fool! But it wasn't true, and I was drinking more and more, and I was encouraging Kate to drink so that I didn't feel as though I was drinking on my own. In hindsight, it was all fucking bonkers!

I mean, why consume a chemical medication in the first place, then put your wine through an extraction process all in order not to get too drunk? Just don't fucking drink surely, for fuck's sake!

It all got very messy over the Christmas and New Year's period. Kate and I were having lots of arguments, sparked by a mixture of her own psychological traumas and anxiety, and my continued drinking. I'd been trying to help Kate with her traumas since I'd met her, and now, resigned to the fact that I could not, and probably should not, I'd been talking to her about her seeing a psychologist. I knew how difficult it was for her to talk about her past, and so we agreed that if she saw a psychologist... then I would go to AA.

I'd already learned in those 20 years that I couldn't blame my drinking on anything other than the fact that I was an addict. I'd been through all the excuses, and some of the main reasons why I believed that I drank were extremely compelling, but I'd come to the point where I'd accepted responsibility for myself, and I no longer relied upon the excuses. I drank because I drank. It provided me with the temporary and perceived feeling of pain relief, relaxation, escapism, and pleasure. Of that, I was completely aware.

I stopped drinking on the 8th of January 2017 and attended my first AA meeting on the 9th. This time it was different. This time it was very clear. Every time I'd stopped drinking before I'd felt the need to remove all evidence of alcohol from the house; this time I didn't need to. We still had alcohol in the house, and it didn't bother me this time. I felt different and more confident that this is what I wanted.

What I was to understand so acutely and profoundly at my first AA meeting, and subsequent research, was that I had a debilitating disease called alcoholism, which whenever I consumed any amount of alcohol, would render me utterly dependent and unable to control the persistent cravings for more. It would destroy my personality, rational thought, and life in general, and would end up killing me. I learned at that time that I would never get over this, or be able to learn to moderate, or be a normal social drinker. Ever. End of story!

That was my spiritual moment of clarity. My 'God' moment!

A Theory of Everything

It provided closure for me in something that had challenged me for so long, and brought me a sense of serenity and peace. And I was, and am, so happy to finally find myself in this position, because I felt that whatever happened from then on, I was in a significantly better position to understand it, to handle it and to maintain my sobriety, and to have established a fundamental and personal understanding of an addiction, and what it took to relinquish it.

Although what I'd learned about the true extent of the 'disease' and its perverse perception within society, utterly disgusted, angered, infuriated and frustrated me, I felt completely relieved and ready for closure on something that had been affecting me for the previous two decades.

Something else I learned was the fundamentals of the process of the 12 Steps employed by AA, and other associated addiction help organisations, such as CA, NA, MA and SLAA. These were life steps I'd already established, in a similar sense, in my quest for personal development and self-awareness, and I'd already, in my own way, addressed all steps; except one step! And that was step one, acknowledging and accepting complete powerlessness over alcohol. That's not to say that I didn't feel as though I wouldn't have benefited from following the steps in order, so as to gain the full benefit, which is what I subsequently did, but it did give me a good feeling that I'd been on the right tracks for the past decade or so already!

Please note that although the references given below are discussing the 12 Steps with regard to alcoholism, they also apply to other addictions, and to life in general. As I said, I found that I'd already established and implemented many of them from my research and understanding of life in general, and they are an exceptionally powerful and useful resource:

Reference: The Twelve Steps of Alcoholics Anonymous (tinyurl.com/ybtkmape)

A Theory of Everything

Reference: The 12 Steps of AA Explained (tinyurl.com/yco3ms8v)

I saw similar tangents for the explanation provided by AA surrounding alcoholism, with the analogy presented by Schrödinger's cat, in a quantum physics scenario. The Schrödinger's cat analogy was developed in order to try to explain the basic principle of quantum physics to the general public and scholars alike. In quantum physics the phenomenon of 'duality' was being explained, where light can exist in the form of either a particle, or a light wave, and its form can only be determined by observation, at the point of which it becomes fixed as either a particle or a light wave, as a result of the observation. So, in the analogy, the cat is locked in a box (with a radioactive source), and the observer determines whether the cat is alive or dead (as it could exist in both states) by opening the box and making the observation.

My point is that it would appear that the explanation provided by AA suggests a scenario where any human being will not establish whether they are part of the approximately 15–20% of the population suffering from alcoholism until they take their first drink, at which point, if they are an alcoholic, then they are already hooked, and on the slippery slope to a miserable life of self-destruction. I subsequently learned that this was not actually the case, but was grateful for the explanation provided at the time, as it enabled me to get to where I am now.

Even though Kate and I separated that February, I'm so grateful to her for giving me the chance to get the information that I needed to work things out, and for providing me with the opportunity to finally understand who I was, and how I could begin to live the life that I wanted to live.

5.2 Projection and Mirroring

Another thing I learned in my time with Kate were the concepts of projection and mirroring. As we argued, there would always be a lot of finger-pointing and blame-shifting. I'd feel hurt by what at the time I perceived Kate 'did' to me, until I managed to see it for what it really was. She was simply mirroring and projecting some of my own flaws.

Allow me to explain. It's understood that what you blame others for, or how you judge them, criticise them, or the 'faults' that you see in them, is often nothing more than a reflection of what's going on within yourself. You'll find this to be true once you start to observe your own negative internal comments about others' behaviours or traits, and see that you may actually display those 'bad' traits or behaviours yourself, or that you'd actually like to have more of such traits yourself. Mirroring your own behaviour in others is called 'projection' and is a very interesting phenomenon to look into!

To give you an example, a friend of mine shared with me how she always got so irritated by her boyfriend's incessant talking to anyone who would listen, not giving her an opportunity to say something herself. She then also admitted to me that she secretly wished she would have the confidence to talk in such a way herself. So, she was angry at him for having the trait she so desperately wanted. This is just one of many examples of projection and mirroring.

So, back then, instead of feeling bitter and resentful, I chose to be eternally and sincerely grateful to Kate for the experience, and the mirror that our relationship provided me with, and more importantly the invaluable gift of finally understanding my relationship with alcohol, which she enabled me to pursue and conquer.

"The most dangerous psychological mistake is the projection of the shadow on to others: this is the root of almost all conflicts"
Carl Jung

"You can tell more about a person by what he says about others than you can by what others say about him" Audrey Hepburn

---

NEAT LIFE HACK: YES, YOU REALLY CAN **CHOOSE** HOW TO THINK AND SUBSEQUENTLY FEEL ABOUT A SITUATION. IT TAKES PRACTICE, FOR SURE, BUT IN MY OPINION, THERE IS NO VALUE TO ANYONE IN CHOOSING TO FEEL RESENTFUL, ANGRY, DISAPPOINTED, BITTER, REVENGEFUL, JEALOUS, OR ANY OF THE OTHER NEGATIVE EMOTIONS THAT YOU WOULD OTHERWISE LET YOURSELF EXPERIENCE. TRUST THAT THERE IS POSITIVE TO BE FOUND IN ANY EXPERIENCE YOU ENDURE, AND ACTIVELY FIND THE POSITIVE

5.3 You Cannot Change Anyone

One of the things I realised very early on from my study of human psychology, and subsequent learning of the deep subconscious, is that *you can't directly change anyone.* If anyone is going to change, it has to be as a direct result of them being aware of the thoughts, emotions and subsequent behaviour that they have and display, and them wanting to change for themselves, and not for anybody else. Even if they do want to change for someone else, it's clear that in the majority of cases it inevitably leads to much disappointment, because they are trying to do it for what I believe to be the wrong reason. Equally, somebody cannot change you!

And it's exactly analogous to one of my primary early beliefs. If you do not truly love yourself, then you are going to struggle to truly love another. If you do not truly understand yourself, then you will struggle to truly understand another. And if you do not truly care for yourself, then you cannot truly care for another.

This sounds selfish on the face of it, but believe me, it's not at all. In my opinion, it's the only way to be able to experience and reciprocate love and compassion and care at the highest and most synchronous level, with anyone who you resonate with, and the Universe itself.

It's exactly synonymous with the example whereby, on an aircraft, they always insist that you secure your own oxygen mask first, before attempting to help anyone else, including, and explicitly before, you help your children.

NEAT LIFE HACKS: AT A FUNDAMENTAL LEVEL YOU **MUST** BE 'SELFISH' WITH YOURSELF INITIALLY, IN ORDER TO BE ABLE TO ENJOY THE BEAUTY AND WONDER OF TRULY GIVING TO OTHERS, CARING FOR OTHERS AND LOVING OTHERS IN THE FUTURE. LOVE YOURSELF BEFORE YOU LOVE ANYTHING OR ANYONE ELSE, AND THAT LOVE WILL RADIATE HARMONIOUSLY OUTWARDS

5.4 Trust and Integrity

One of the things that I was very grateful for in my relationship with Kate is for my beginning to think about trust and integrity, and my investigation and basic understanding of anxiety.

I've always been acutely conscious of my desire for trust and integrity, despite my lies and lack of integrity as a child (and young adult). I've always felt far more comfortable when I've allowed myself to be honest. I've always prided myself for wearing my heart on my sleeve. I'm not afraid to tell people what I really think. This isn't always productive, as I'm sure you can imagine, but in general, it's well received. Also, I think that as a result of my now developed natural honesty, people are more inclined to trust me more quickly, and share their honesty and openness with me in a similar fashion.

"Being honest may not get you many friends, but it will get you the right ones" John Lennon

---

But there's a fine line, and it took me a long time to establish barriers, and to work out whom I could trust, and whom I couldn't, or shouldn't. I'm naturally giving and accommodating, and have always enjoyed helping people where I can. So, my policy evolved into one where I would trust someone immediately, until the point where they let me down, and then I would gently disengage with them, and remove them from my life.

After reading a great book called *The Road Less Travelled* by M. Scott Peck, I began to put my understanding of human psychology into better perspective. I was to have highlighted that it's second nature for human beings to lie and be untruthful: *a lot*! In fact, these 'skills' are beginning to be developed at your age and earlier, and will be utilised by you throughout your entire life, for a varying level of reasons and purposes. The underlying reason provided for our propensity to lie is that in certain situations it's simply easier to get things by lying and deceiving, than it is to get things through violence and brute force.

Book Reference: The Road Less Travelled – M. Scott Peck (tinyurl.com/ya3o2e38)

So, whilst I accept that learning to lie and deceive people appears to be part of 'modern' society, and apparently a way of getting what you want without having to resort to violence, I'd suggest that there are far better and more harmonious ways of achieving what you want than having to be untruthful. You see, the very action of lying and deceiving lowers your vibration, and in fact, when you are honest and integral, then you raise your vibration, and, as a result, you are attracting more of what you want into your life, even if it doesn't come immediately from the source that you're expecting it to come from.

5.5 Assumption and Presumption (and Speculation)

So, lying and dishonesty are one thing, but there's another part of human psychology which I consider far more destructive and damaging, and that is *speculation* surrounding **assumption**, or **presumption**, or even just speculation itself. Assumption is making a judgement on something based on a guess, or hearsay. Presumption is making a judgement on something based on probability, or only a little information. Presumption is better, but both are *pure evil* in my opinion.

I feel that nobody should present themselves as an authority on anything at all, without having direct experience of it themselves, unless otherwise quoted as a reference.

What I mean is, before I describe my understanding of anything to anyone, I ask myself, 'Do I really have enough information and experience to tell this story like I know it to be true?' 'Am I speculating or making any assumptions or presumptions?' If the answer to any of those questions suggests that I am not in a position of authority on the topic, then I will make sure that I state this clearly in my dialogue. If it's important, then I will subsequently do some research and put myself in a better position of authority.

NEAT LIFE HACK: WHEN COMMUNICATING WITH OTHERS, IN PARTICULAR THOSE YOU RESPECT, ASK YOURSELF WHETHER YOU ARE SURE THAT WHAT YOU ARE SAYING IS TRUE, OR WHETHER YOU ARE SIMPLY REGURGITATING HEARSAY OR RUMOUR, AND ADJUST YOUR DIALOGUE ACCORDINGLY. IF YOU DO THIS, THEN

A Theory of Everything

So, going back to speculation, assumption and presumption: what a challenging situation!

It would appear to me that a large proportion of society is very comfortable operating and making their judgements of people, events, and situations, based on speculation and assumption alone, and usually without a consideration to the fact that they might be, and often probably are, completely and utterly wrong! That's insane, isn't it?!

I think that this is an exceptionally destructive and damaging social situation, but I have to accept it, and I had to work out how I was going to deal with it, and how to make the best of the situation. So, for me this looked multifaceted.

Firstly, I immediately further developed my bullshit detector.

Then I made a deal with myself that I wouldn't feel any guilt for not wanting to, or being unable to help people, and for removing myself from the presence of people who are simply not in a position to be helped, for whatever reason.

I do not blame anyone for anything, or wish to judge anyone. I'm pretty sure that we are all as misaligned as we are in general, as a result of our childhood traumas and society, and the system put in place to keep us dumb. So I accept that it's not their fault; or my fault. I do, however, hold everybody responsible (including myself), which is a very different sentiment.

"I don't think I'm smarter than anyone, I just see things that others don't see. I'm not here to judge, I'm here to point out things you may or may not understand, and make you think" Keen James

---

A Theory of Everything

Yeah, that's kind of where I'm coming from.

NEAT LIFE HACK: TRUST EVERYONE WITHIN REASON IMMEDIATELY UNTIL SUCH A POINT THAT THEY LET YOU DOWN, AND IF THIS HAPPENS THEN LET THEM DOWN GENTLY AND REMOVE THEM FROM YOUR LIFE

"Love everyone you meet from the moment you meet them. Most people will be lovely and love you back, and you can achieve the most wonderful things. But get rid of any of the b@stards that let you down" Joanna Lumley

---

See, I learned that effective communication, at a deep level, is far easier when your energies are resonating at the same frequency (hence the expression, 'I really resonate with him/her'), and that many people are simply not ready to be helped, or need a different type of help than you want to provide, or can provide. It's often the same as addiction. You can only want or receive help when you are critically aware of the help that you need, and you accept that you need it. Only then are you open to accept, and receptive to the advice of others.

"There are some people who could hear you speak a thousand words, and still not understand you. And there are others who will understand without you even speaking a word" Unknown

---

5.6 Raising Your Own Vibrational Frequency

Here is quite a good summary of the basics surrounding raising or lowering your vibrational frequency:

"Vibrational Frequency:

Raising your vibration – Gratitude, kindness, love, joy, passion, forgiveness, acceptance, sunshine, walking in nature (particularly barefooted – grounding), breathing deeply, yoga, meditation, exercise, laughing, smiling, hugging, singing, dancing, raw wholefoods, greens, fruits, nuts, creativity, relaxing music.

Lowering your vibration – Junk food, alcohol, toxic relationships, negative thoughts, environmental toxins, toxic products, excess red meat, white sugar and sweets, medication, radiation, yelling, arguing, holding onto the past, anger, resentment, guilt"
Unknown

PART 6
My Ultimate Learnings

6.1 Significance of Our Thoughts
– It All Starts in the Womb

One of the primary focal points of my learning journey had been on the fundamentals of how our minds work, and I learned why it is that we sometimes behave differently than we think we should, or even differently to how we consciously believe that we actually do.

You see, it all seems to start when you're in the womb.

Yes, even before you're born, you're absorbing information at a phenomenal rate, like a sponge. So, up until the age of about two, the memories of whatever you've felt, heard, or experienced, especially on an emotional level, become pretty much your '**core belief blueprint**' for the rest of your life. This blueprint will dictate your thoughts, emotions and subsequent behaviours. Unless you seek to find a way of addressing this later in life.

This period from womb to the age of two is referred to as the stage where your brain is vibrating at a much lower frequency than it does when you're an adult. It is said to be in Delta brainwave state, or, the lowest brainwave cycle of 0.5 to 4 cycles per second.

A Theory of Everything

Between the ages of around two to ten, you're said to move through Theta (4 to 8 cycles per second between the ages of about two and six) into Alpha brainwave state (8 to 13 cycles per second between ages of about five to eight). In these states, your brain is vibrating at a higher frequency, but you're still absorbing information from what you hear, and the experiences that you have. It's just at a much slower rate than when you were in Delta brainwave state.

At approximately the age of ten, although your brain is still developing, and will continue to do so until about the age of 22, you progress into Beta brainwave state (13 cycles per second or above from ages eight to twelve onwards), and that's it, **boom**! Your **subconscious** memory is locked down tighter than Fort Knox! Anything and everything that you've seen, heard, felt, or experienced in this brief period of your life, now subconsciously plays a big part in how you behave, as well as your general outlook and approach to life. And it is this which appears to be partly responsible for your *undesirable* traits, and what causes you to act/react the way you do, which is not the real or pure *you*.

Reference: Understanding the brainwaves of your children (tinyurl.com/y8khn9vk)
Reference: Bill Harris – Centerpointe – AudioBeats

Of course, there are other influences that affect your behaviour, like so-called genetically inherited traits, your DNA (which I have learned can be changed through conscious thought), as well as the date and time when you were born and your birth number (astrology and numerology, respectively), and past life traumas, for example, but I think that these traits are predominantly more to do with the '*true* you', rather than the 'preprogrammed' and distorted you.

It seems apparent that a trauma or shock at any stage in someone's life can affect their deep subconscious and core beliefs, and change the way they view themselves as well as the way they behave or react to situations. But once you learn to address your subconscious beliefs, you can then regain control. And with that find the happiness that is ultimately available to you and, indeed, is the *true you*.

"The privilege of a lifetime is to become who you truly are" Carl Jung

---

So, let me try to put this into perspective. Pretty much everyone goes through some hardship or traumatic experience at least once in their lifetime, and most importantly, from when they are in their mother's womb, to before they reach the age of ten. A good example is growing up in a poor versus a rich environment.

Imagine having been brought up into a poor environment which focussed mostly on being 'poor'. From age 0 up to age 22 your environment was constantly affirming to you that life is financially challenging, one of hardship and that the things you want can't be afforded. These messages then became deeply rooted in your subconscious, and become your core beliefs and expectations of life. Unless you decided to make the time and effort to acknowledge these beliefs, accepted them, and attempted to address them.

Similarly, if, while you were growing up, the messages you received from your close environment were ones of abundance and wealth, then in turn, your core beliefs would be that abundance and wealth are readily available and flow into your life easily.

This could perhaps also explain why children from violent and abusive families tend to struggle with these same behaviour traits as they grow up. Of course, there are exceptions to this rule. Some poor families are perfectly happy and balanced, and raise their kids with a better head start in life as a result. These families are not financially rich, but they are emotionally and spiritually wealthy, and, despite their often significant challenges, endowed with much happiness as a result. And from these poor communities can often rise a star, like a phoenix from the ashes.

To try to put this into context, I find it interesting to consider civilisations who apparently live in significantly better harmony than we do in our 'modern' societies, and whose children are born into more stable and healthy environments, like many indigenous peoples such as the Native Americans or the Aboriginals. The Amish community, for example, is rumoured to have **no** recorded homicides in their entire history. No major crime, no murders, no jails, no institutions, no social out-casting, no intolerance, no prejudice, no racism, more happiness, more freedom, more understanding, and healthier and more harmonious lives.

So, how can all of this help you? Well, primarily, I'm hoping it'll plant the seeds for you to at least become **aware** of your thoughts and behaviour at an early age.

Please see Appendix 1 for an overview describing much of what I have discussed and presented, taken from Zeitgeist III. (youtu.be/16wdfA8CNdY)

I learned the significant importance of becoming aware of my behaviour and thoughts, and the difference between how I perceived myself, and how I actually behaved. I learned that how I thought that I *should* behave was more akin to my true self, and how I *actually* behaved was a result of the profound experiences that I had had in this lifetime, and others, which were deeply seeded in my psyche, and causing the blockages that I'd been endeavouring to work with.

I learned about the instantaneous interaction between our conscious mind and our subconscious mind, and the roles that they play in structuring and formulating our thoughts, and dictating our subsequent beliefs, which result in the behaviour that we display.

I learned that our conscious mind operates in the present, and is influenced by our intuition, and that it's not responsible for the information that is used to consider how we should perceive or react to a situation.

I learned how our subconscious mind operates in the past, and entirely from memory, and that it will more commonly default to the most negative memory, as a mechanism to protect ourselves from an environment that has usually changed. It is entirely responsible for the information and beliefs that result in our outlook on life and subsequent behaviour that we display.

I learned how our conscious mind has a tendency to construct future scenarios based on the information that it receives from our subconscious mind, and how I could begin to positively influence the thoughts that I had, and how this could positively change the perception that I had for the future, and subsequently positively effect what I attracted to myself.

6.2 How to Change Your Social Blueprint, Core Beliefs and Subsequent Behaviour

I learned how and where to remove negative thoughts, and in doing so, I realised the greater clarity this provided me with, and the positive impact that this had on my life in general.

I realised how powerful this was in positively influencing the relationships and interactions that I had with everyone and everything that I engaged with, and how I could raise my vibration and how I could influence and make positive what I attracted to myself and into my life.

And I realised that in doing so, I could open myself more freely, and with far less negative bias, to the information coming from my intuition and user guide, and could make far wiser choices and better decisions, which positively influenced everything in my life from that day onwards.

A Theory of Everything

I learned about how we can begin to *choose* to perceive our thoughts positively, and how to manage and find clarity in our thinking. I learned that you could do this by observing your thoughts as they arise, and how you could remove redundant negative thoughts and associated emotions immediately, leaving you with far better clarity with which to consider your position and subsequent actions and behaviour. The schematic below aims to explain the basic process:

Schematic 1 – Simple diagram showing how negative thoughts affect our behaviour, and how to work with them

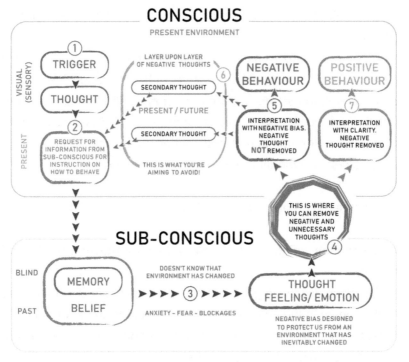

Simply put, you have two types of consciousness: the **conscious mind**, and the **subconscious mind**, as shown in the schematic above. It is of course slightly more complicated than this. For example, the subconscious also

has recall memory, and pre-recall memory, and there are also memories associated with past lives, but let's keep it simple for now.

To recap, our subconscious **beliefs**, which are derived from memories, act as a primary reference for our conscious mind, to determine immediately how to perceive a situation, and how to behave as a result. Our conscious mind requests information derived from **memories** stored in our subconscious, and our conscious interpretation of that information dictates our **behaviour**, or how we respond to a situation.

Your **conscious** mind represented in the top of the schematic operates in the **present**, and it's where your thoughts are triggered. It's **sensory** (through sight, touch, smell, hearing, pain, etc.), and it is not responsible for information.

Your **subconscious** mind, represented in the bottom section of the schematic, is your information storage hub. It's **blind**, and deals in thoughts and emotions derived from memories, often linked to traumas (or elations) from the past. This is where your beliefs and associated blockages are stored, often deeply buried under layer upon layer of secondary negative thoughts and associated emotions.

So, in my interpretation, and with reference to Schematic 1:

1. Your **conscious** mind will be presented with a **thought**, produced by a **trigger**
   - For example, the trigger can be something that it sees or hears happening in the now, or a sight, smell, touch, sound or interaction with a person, which requires some action to be taken, or some further consideration to be made
2. Your conscious mind immediately defaults to information derived from **memory** stored in the **subconscious** mind, for explicit instruction on how to subsequently think and **feel** (**emotion**) about the situation, and

how to respond to the situation, or how to act or **behave**

- In the human psyche, this would appear to be more often a negative thought, coming from a memory, which forms your true core belief. The reason that it's predominantly negative is that it's designed to protect you from your environment, **even when your environment has changed**

Frankly, I think it's an imperfect and outdated system, and only serves us good purpose early in life, but then appears to hinder us substantially for the rest of our lives. Ironically, a price we seem to have to pay for the best possible chance of staying alive in the early stages of our life!

3. Because your subconscious mind is blind and operates in the past, and is not sensory, and cannot actually see (hear, smell, touch, etc) the present situation, it will provide you immediately with a thought associated with your most prominent **memory** and associated **emotion** of what happened in a similar situation in the past, and commonly from your early childhood

- This thought and associated emotion takes precedence over any other thoughts, and if negative, the more profound the trauma associated with it, or the more often a certain negative scenario repeated itself, the deeper it will be buried, and the more challenging it will be to identify, or change, and the more layers of negative thoughts will have been built around it, and on top of it

- You will know whether this thought is negative or positive quite clearly because of the way it will make you feel. **If it makes you feel good, then it is a positive thought, and if it makes you feel bad, then it is a negative thought.**

- These negative memories are often related to fear, anxiety and blockages

4. We're in the yellow star now. This is where it gets interesting as this is where you'll be able to become aware of, and identify, your negative thoughts, memories and subsequent beliefs. This is where you'll be able to make a positive impact on your perspective and life, and everything that you experience and that happens to you
   • When your thought in 3 is positive, and it makes you feel good, then it runs straight through 4 into 7, and your subsequent behaviour is positive
   • Please see sections 6.5 to 6.9 for details of how to do this

5. But let's look at what often happens, before you've trained yourself to be aware of the polarity of your thoughts and associated emotions:
   • When you experience a negative thought as a result of a trigger, which has come from a negative memory, then your interpretation is going to be equally negative, and this results in negative or undesirable behaviour, whether you're conscious of it or not

6. A common process observed in the human **interpretation** process is to create further thoughts surrounding the original negative thought, which subsequently produce more **secondary negative emotions that strengthen or bury deeper the initial negative thought**
   • These negative thoughts can be associated with the present, but are more often than not constructed surrounding future (made-up) scenarios. These are usually utterly unfounded and completely untrue, and based upon our own misinterpretation, speculation, assumption and presumption, rather than on the actual facts, and as a result of the lack of clarity experienced by indulging the negative thoughts
   • 6 moves back into 2 and through 3 and 4, and back into 6. These are the layers. And they can often build up so much that they cause unnecessary anxiety and depression as they spiral out of control and

produce unnatural and erratic behaviour. As I said, these thoughts are usually seeded in the future stemming from anxieties and fears, creating negative future scenarios, which are simply not true. However, the more you think about them and convince yourself, the more likely you'll attract them to yourself, materialise them and experience them

- Please see Schematic 2 and 3 for direct examples, as explained below

7. So, the key is to train yourself to become aware of your subconscious thoughts as they are evoked by the conscious triggers, and to remove them and to change them to be positive. I call this silence of the Monkeys!

Please see schematics 2 and 3 for examples, which describe a couple of specific situations. They show what might normally happen, and what you're aiming to avoid by using this technique.

Schematic 2 shows an example where you've had a disagreement with a work colleague or your boss. It aims to describe the normal thought-process whereby layers and layers of negative thoughts are built up, which are derived from negative thoughts stored in your subconscious.

It describes what you are going to be able to avoid by applying this process using the techniques for removing negative thoughts (and their layers), which are given in sections 6.5 to 6.9. I've found all of these techniques to be equally powerful; however, I have found that I most commonly default to the method described in 6.9, as I have honed it to be able to perform it instantly. Until I was able to get to that stage, I used one, or a combination of the others in the interim, and I will often use any of the others as and where I feel appropriate.

I'd suggest that you play around with any of them, you find your own preference, and your own method or technique that works for you. In my experience, there are no particular rules for how you construct the dialogues or visualisations in your mind, just remember, and have faith in the fact that your mind is an exceptionally powerful device, once you have worked out how to use it to its best advantage! Also remember that you will know when the techniques that you are using are working, because you will tangibly observe the sensation of physically relaxing.

Schematic 3 shows the same process based on an example where you've had an argument with your partner or friend. The same principles apply.

A Theory of Everything

Schematic 2: Example applying the technique to where you have had a disagreement with a work colleague, or your boss

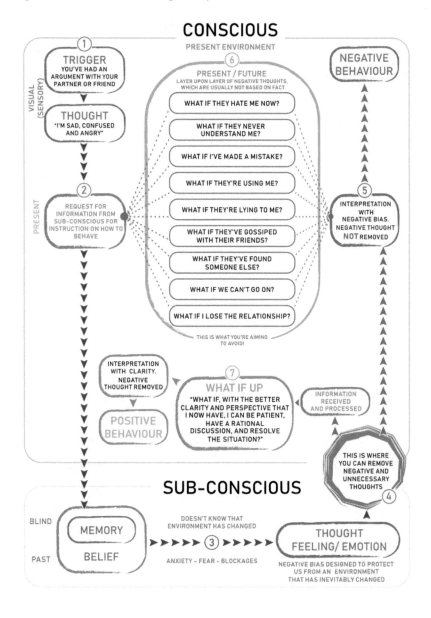

Schematic 3: Example applying the technique to where you have had an argument with your partner or friend

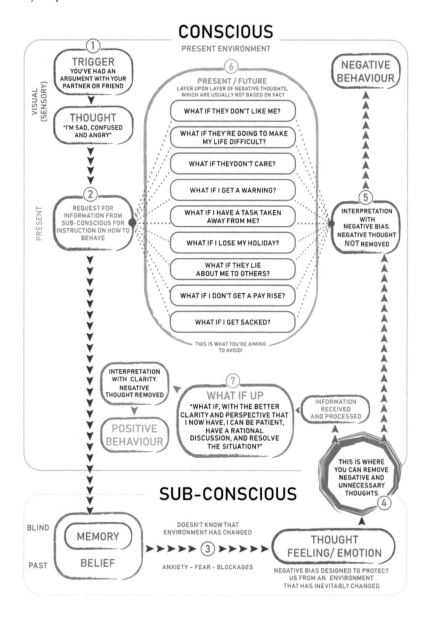

6.3 Silence of the Monkeys

If you want to change the way you're feeling or behaving, then one way of doing it is to *simply* become **aware** of your *negative* **thoughts** as they're being evoked by your conscious triggers and brought up by the subconscious, and to remove those negative thoughts, in order to enable you to indulge in that particular situation with clarity, and in a positive light. Essentially, you have to declutter your thought-process. This is a process that I have called *Silence of the Monkeys* or *Monkey Hunt* stemming from the commonly observed phenomenon of 'Monkey Chatter' or 'Monkey Mind'. Please don't confuse my work with that of Prof Steve Peters, *The Chimp Paradox*. This is a good book, and I can see how it has been so effective for a great many people, but it's not for me, and has not influenced the work that I'm presenting here.

Book Reference: The Chimp Paradox: Professor Steve Peters (tinyurl.com/ydbll79e)

I say *simply*, but this is of course quite challenging. As I've mentioned, the human psyche defaults to the most negative memory as a tool for protecting us from danger. Moreover, society also programmes us to be negative, fills us with fear, hatred, prejudice and anger, which is utterly unconducive to a healthy and happy life, and a positively functioning mind. Equally, since we tend to be raised in unconsciously negative environments, and experience undesirable traumas in early life, even as we reside in the womb, then the associated blockages, fears and anxieties can be challenging to address, and will take time to work on and to fully clear.

A Theory of Everything

"You have to look deeper, way below the anger, the hurt, the hate, the jealousy, the self-pity, way down deeper where the dreams lie. Find the dream. It's the pursuit of the dream that heals you" Lakota prayer

So I learned that if you become **conscious** of your **thoughts** as they happen, and gradually replace the default negative memories (or programmes) with positive ones, then you will see remarkable changes. Not only in how you view the world and your experience, but also in the things that happen to you, and around you. Like attracts like. Positivity attracts positivity. You can only send out either a positive vibe or a negative one to the universe, and with careful analysis, you can begin to choose which vibe you're sending, and what you're attracting back into your life.

In doing this, I also realised that I was engaging with other people in a far more positive light, and in the absence of self-sabotage resulting from the negative and often wrong perspective that I was harbouring and projecting from my redundant subconscious thoughts and beliefs. I accept that in the past I've ruined relationships by superimposing my own negative memories and thoughts on my perception and expectations of others, and as a result, I've forward-projected a negative expected outcome, which I've subsequently made myself open to, and attracted to myself. I've seen that this has been unnecessary and I'm very happy that I'm able to recognise it and do something about it in the present. I've observed myself transform my outlook towards others with dramatic effects on the subsequent relationships that I've built.

This also forms one of the steps I've taken towards becoming Unfuckwithable! (tinyurl.com/y7mhfsdq)

6.4 Stress – It's Your Choice

It's understood that stress is by far one of the biggest killers in life.

But I've learned that you have a **choice**. You can *choose* to get angry, or not. You can *choose* to let things stress you, or not. You can *choose* to let things piss you off, or not. And you can choose to use techniques to combat stressful and negative thoughts.

And you know that I haven't found this easy, and that I still find it challenging sometimes; but I am so grateful to you for providing me with the best reason that I could possibly have wished for, to be aware of the beliefs (memories) that I have, which influence the thoughts that I have, by providing the emotions that I have, which produce the behaviour that I display.

I can't go back and change the past, but I can do my best from now on to be a better father. Here are some techniques which have helped me greatly to handle on my own stress levels:

6.5 Simple Process for Managing Emotions I – Acknowledging, Accepting and Allowing

One way to dispel negative thoughts is to hold them in your mind and not allow them to evoke further negative thoughts, which otherwise lead to a downward spiral and layers and layers of further negativity, fear and doubt. An effective method for doing this is to consciously acknowledge the thought, to accept it, and to allow it. To make it OK.

This is a challenging process, and I'm still catching myself having negative thoughts every day, but I can assure you, the more you practise, the easier it becomes, and the quicker you begin to become aware of them.

Acknowledging them, *Accepting* them, and *Allowing* them, is key. If nothing else, it puts you in a better and more positive position to deal with your challenges as they occur.

A Theory of Everything

You can look at it as if the primary purpose of a negative thought/emotion is to disrupt you and cause you to feel negative. For example, resentful, angry, bitter, confused, jealous, hateful, spiteful, regretful, revengeful, or in a way that you don't understand but are not experiencing as positive. An easy way to immediately recognise the difference between negative and positive emotions is by observing how the thought/memory/belief makes you feel. Quite simply, if it evokes a bad feeling, then it's inherently negative, and if it makes you feel good, then it's positive, and doesn't need addressing!

It can be viewed as if negative thoughts/emotions really want to beat you up! Left to their own devices, they will nag at you, and nag at you, breeding further negative thoughts and triggering negative emotions, until your thoughts, emotions and feelings spiral exponentially downward to the point where you might even break.

But guess what? If you *acknowledge* them, *allow* them and *accept* them, and make them *OK*, then they've got no *fuel* to sustain their existence. They've got nowhere to go. And so, with practice… they simply vanish, as if they had not ever been there in the first place, and once removed, they're permanently gone.

Yes, yes, for sure there'll be residual thoughts (or layers) stemming from the one you just removed, but if you use the same process on them, then you will soon have a stiller and clearer mind to deal with the situation that you need to deal with, and now in your best capacity.

And yes, sometimes negative thoughts can come into your mind for a distinct purpose, and because you **need** to consider them. I've found when I practise what I've just described, and remove the negative thoughts and associated emotions, then this allows me the clarity to think about the situation constructively, and without negative bias, and in the absence of unnecessary speculation, blockage, distraction, doubt or destruction. I have silenced my monkeys!

NEAT LIFE HACKS: IF YOU EXPERIENCE A NEGATIVE THOUGHT (STEMMING FROM A SUBCONSCIOUS BELIEF, OR MEMORY AND ASSOCIATED FEELING OR EMOTION) AT ANY TIME, YOU WILL FIND THAT IF YOU CONSCIOUSLY ACKNOWLEDGE IT, ACCEPT IT, AND ALLOW IT, THEN IT HAS NO PLACE TO GO, AND IT WILL SIMPLY DISAPPEAR. IT'S A BIT LIKE WHEN SOMEONE IS ANNOYING YOU, ANTAGONISING YOU, OR WINDING YOU UP AND YOU JUST ACCEPT THAT THEY ARE THERE BUT DON'T TAKE THEIR BAIT, AND DON'T RISE TO THE SITUATION, THEY WILL GET BORED AND EVENTUALLY LEAVE YOU ALONE

6.6 Simple Process for Managing Emotions II – I Love You

Another one of the methods I learned, which Chris New taught me early on for this purpose was, whenever you have a negative thought in your mind, which you're not comfortable with, and which will probably lead to further negative thoughts, if left to its own devices, then simply repeat a mantra in your mind: '*I love you, I love you, I love you*, etc.'. In this way, I've found that I can very easily and quickly remove negative thoughts, which serve me no purpose, that would otherwise consume me and lead to fear, anxiety and worrying, and subsequently attract that negativity into my life.

NEAT LIFE HACKS: IF A NEGATIVE THOUGHT COMES INTO YOUR MIND AND YOU WANT TO ERADICATE IT, THEN SIMPLY REPEAT THE WORDS 'I LOVE YOU' IN YOUR MIND, UNTIL THE NEGATIVE THOUGHT HAS VANISHED

6.7 Simple Process for Managing Emotions III – The What If 'Up' Technique

This originated from a Joe Vitale presentation

NEAT LIFE HACKS: THE **WHAT IF 'UP'** TECHNIQUE. THIS IS A VERY NEAT AND SIMPLE TECHNIQUE TO HELP YOU WITH NEGATIVE THOUGHTS AND RESULTING EMOTIONS. FOR EXAMPLE, CATCH YOURSELF THINKING 'WHAT IF THE WEATHER IS GOING TO BE TERRIBLE FOR THE MATCH TOMORROW?' AND CHANGE THAT THOUGHT TO 'WHAT IF THE WEATHER IS GOING TO BE A SCORCHER TOMORROW, AND WE HAVE A LOVELY DAY?', OR 'WHAT IF, WHATEVER THE WEATHER DOES, I'M GOING TO BE HAPPY ANYWAY'. IT'S FREE, AND YOU'VE GOT NOTHING TO LOSE, AND IF IT DOESN'T MAKE YOU INSTANTLY HAPPIER, THEN I'LL GIVE YOU YOUR MONEY BACK!

6.8 Simple Energy Technique to Remove Stress, Anxiety, and Worrying – Part I

Stress is often caused by negative energy. Negative energy from others, or from a situation you're in, or from thoughts within you, as discussed previously. Something important I learned was a technique to protect yourself from negative energy, in situations where it's otherwise unavoidable, or any time at all that you feel threatened by negativity or the negative influence of others.

First of all, let me explain my understanding of the term, 'the source'. By this I mean Universal Energy in its most divine form. Love, unconditionally and all-encompassing. I've come across it, as you will have done, too, in many different religions, concepts, ideologies and practices, but in my opinion, it's all the same thing, and it doesn't matter how you describe it, or what you attribute it to. It's the same. So, whether you call it 'God', 'Allah', 'Buddha', Gitche Manitou, Krishna, Vishnu, or whatever you believe in, it doesn't matter, as long as it feels right with respect to your own belief.

My own personal belief is that **God *is* Love**. Or put another way, ***Love* is God**.

A Theory of Everything

In the full process, I will take three deep breaths and ask to be connected to 'the source', and then I will ask my Guardian Angel to put a six-foot barrier up around me to protect me from evil, and I visualise myself being surrounded by that light, and I see the force field going up (rather like a reinforced aura).

Often, I will feel myself physically relax immediately, in a similar sense as when I remove negative thoughts, as described in 6.9.

If this is not the case, then I've learned that it's likely that the negative energy is actually coming from within me, and not from an external source, in which case I ask for the negative energy to be expunged from within the 'force field', and I visualise this happening. Usually, this then removes any residual negative energy, or thoughts, and I will immediately experience the physical sensation of relaxation. I can then get on with thinking about it, and doing things that are more productive, rather than spending my time worrying and stressing.

Honestly, I don't know for sure if my Guardian Angel actually exists, or how the process works, or if it's even just a placebo effect. But you know what? I don't give a damn, because, however it works, it works, and it provides me with what I need to find better clarity in my thoughts, and to be more productive and fruitful as a result! It also reduces my stress levels and anxiety immediately and completely.

NEAT LIFE HACK: WHEN YOU FEEL THAT YOU ARE THREATENED BY NEGATIVITY OR NEGATIVE ENERGY, ASK TO BE CONNECTED TO THE SOURCE, THEN ASK YOUR GUARDIAN ANGEL TO PUT A SIX-FOOT FORCE FIELD AROUND YOU TO PROTECT YOU, AND VISUALISE THAT BARRIER GOING UP AROUND YOU.

6.9 Simple Energy Technique to Remove Stress, Anxiety, and Worrying – Part II

The second stress-reducing, energy-healing technique that I'm about to describe changed my life with dramatic effect. It is a method that I have put together from my own research and practice, and much insight from the group that I am a part of at The Wellbeing Clinic, in particular, from Jill Sudbury, Ankara and Ivan Brownrigg. And from your Aunty Hannah who, as you know, is a lightworker. In energy-healing work, a lightworker uses their natural psychic-tuning abilities to see and feel their client's energy centres. The energy work may also incorporate psychic healing, in which you tune into emotions and thoughts that are creating problems for a client.

This method instantly removes any of my negative thoughts or feelings permanently, and I feel a tangible sensation of relaxation immediately.

For me, it removes **_unnecessary stress_**; it removes **_anger_**; it removes **_fear_**; it removes (or clarifies) **_doubt_**; it removes any **_anxiety_**; and it removes any **_unnecessary speculation, assumption or presumption_**. Equally, I've not had a sleepless night as a result of worrying about anything since I developed the technique. I haven't got a clue what to do if I can't sleep because I'm excited about something, but that's a good problem to have.

I can't begin to explain how powerful this process is for me, but I accept the fact that it is _my_ process, and won't necessarily work for you, or for anyone else. I am, however, utterly convinced that if you, or anybody, wanted to develop a similar technique, then you'd be very capable of doing so, with the right understanding, knowledge, practice and experience.

So, what I did in the early stages of development of this process was as follows:

A Theory of Everything

- I take three deep breaths and ask to be connected to the universal source
- I visualise myself, wherever I am, connected to the light source, and see the energy pouring in and out of me by connection to the Universe
- I visualise myself on a boat approaching a beautiful and deserted tropical beach, greeted by smiling, loving, and happy people
- I see myself disembarking and walking up the beach to a beautiful, jewel-encrusted marble arch and hallway
- I pass through the hallway and up a hill to a rainbow bridge carrying the load that I want to expunge from my mind. Sometimes this is a specific load, because I'm acutely conscious of the beliefs, feelings and emotions that I want to free myself from. Other times I don't know what's bothering me. Either way, the technique works
- I consciously take two steps 'down', and then three steps 'up' to the bridge
- When I'm standing on the bridge, I look down over the side, and I consciously throw my trash over the side, and watch it as it disappears into the abyss
- I always, and without fail, physically feel myself relax immediately after I throw my garbage over the bridge. I can continue to throw stuff over the bridge until I feel calm and relaxed
- Immediately after this process I find that all of the thoughts that were causing me to feel bad me have simply evaporated and disappeared, and that I can maintain that state of calmness, and I can think constructively and effectively about what I need to consider, without the burden of fear, anxiety, confusion, unnecessary doubt, or any negative bias
- I then visualise a fountain on the bridge, rather like a drinking fountain, but a fountain of light, where I drink the light and imagine

the positive beliefs that I am affirming to myself, to replace the negative that I've just released. If the negative was specific, then I can focus on specific positive memories, and if not, then I just drink.

As with the first process, I've no idea how or why it actually works, but I don't care, it simply does. And profoundly so, which is all that matters to me! Again, I've considered that the physical relaxation is a psychological effect (placebo), and nothing to do with Angels, or bridges, or fairies, or unicorns, but again, I don't care, because it works!

I've got to the stage now whereby I can remove negative thoughts (memories) as soon as they are triggered by my conscious process, and where necessary, replace them with new positive memories or thoughts, which subsequently enables me to think constructively and clearly about the thought, and provides me with a far more stable, happy and harmonious existence.

The explanation for the effectiveness of the above described techniques may simply be by the power of the placebo effect combined with the power of the mind.

6.10 Placebo is Possibly the Best Man-Made Drug

I have come to understand that the most powerful and effective man-made medicine ever created is the *placebo*. A great many studies have demonstrated the immense power of this *'drug'*. It's amazing what our minds are capable of, when we truly believe!

For example, I read a story recently that described a child living in the Amazon, or somewhere quite remote from 'modern' society, who spent his days playing with nature and his environment. Whilst playing, he noticed how when you pulled the tails off salamanders, the tails would eventually grow back. So, when his leg was bitten off by a crocodile one day, he simply grew it back! Of course, I don't know how true this story is, but I put it into the 'allow and watch with interest' section of my internal hard drive, because I can see how there may well be an element of truth to it. I mean, when you cut yourself, you expect it to heal, right? I believe that the body is an extraordinary device, if you know how to use the power of your thought! Many illnesses can and have been fixed solely by utilising the power of the placebo effect. Here is a very good BBC documentary on it:

Documentary Reference: BBC Documentary – The Power of the Placebo – BBC Horizon (tinyurl.com/ya9pgcm2)

6.11 Tangible Experiences Proving the Power of the Mind

As I'm educated as a scientist, it's been quite important for me to see tangible results from the experiments I do, or the actions that I take based on the control I now feel I have over my thoughts. I was trained to believe that I must be able to test, measure, and quantify anything in order to believe it to be true or worth reporting (or perverting). Over the years I've now been able to gather a number of tangible instances which to me have proved conclusively that the power of the mind can allow you to achieve remarkable things.

For example, being able to take ice-cold showers! When I was at my fattest, physically, and actually sat in a wheelchair often, such as when we were at Center Parcs that time, or when we visited the London museums, I realised that I was bitterly unhappy with my condition. I was acutely aware that it went against all of my life principles and that only *I* could do anything about it. I knew that the one thing I could do that would have the most profound effect on my health and mobility was to lose the weight I'd apparently gained as a direct result of quitting smoking.

A Theory of Everything

One of the measures I took during those six months was to have a cold shower every morning. Because, from what I understood, subjecting yourself to thermal shock forces your body, digestion and metabolism into action, and really kick-starts your day.

But, Jeese Louise! A freezing cold shower?! Really?! Did I really have to subject myself to that? Well, I decided that yes, I did, and, yes, I would. It was horrible, and tantamount to torture! I knew that having a cold shower in the morning would kick-start my metabolism and help to regulate my wellbeing in general. So I developed a technique that not only made this practice more doable, but actually made it enjoyable, and fun.

You see, what I did, before I turned the dial on the shower to freezing cold, was to imagine that I was standing on a tropical beach in the blazing sun, and that I was desperate to be cooled down.

It was as simple as that. It worked. Every time. With those thoughts in mind, when I turned the shower down to freezing cold, I felt and experienced immediate joy and benefit from having done so.

I'd do it three times. Hot to cold, and then back to hot again, and I measured a tangible result in as much as that I could feel my body starting to digest and metabolise. First burp and I was done!

It was the same with swimming in the morning. Both practices for me were done with a similar purpose. Heat followed by cold to generate health benefits is the rationale behind the principle of sitting in a sauna, and then rolling in the snow, as practised in many Scandinavian countries. Dunking in ice-cold water is now even regarded as a well-researched strategy for combatting depression.

Despite all the odds, and what other people told me, I lost 25kg (55lb or 4 stone) in just six months, and I regained the comfort and level of mobility I had previously enjoyed. And I kept the weight off for several years, too. Of course, this wasn't solely due to the cold showers and swimming, but I do believe these helped! I elaborate on the details of my weight loss journey in another book.

Another story, in a similar vein, that I want to tell you about is a technique that Alan's dad, Mike Parsley, developed for keeping warm when he was working in his garage in the winter. When he was freezing cold, to the point that he could no longer work on his cars, because his hands and body were so cold, he would imagine that there was a fire right in front of him keeping him warm. By doing this, he was actually able to raise his body temperature and continue working as a result. He used the power of his mind to enable his hands to work properly again!

This sounds incredible, if you think about it in the way we're taught to believe, but *actually* it is a very credible and a well-documented phenomenon. It's amazing what the mind is capable of, if you understand its principles, and believe in yourself and your abilities!

Part 7
Great Pieces of Advice I Was Given

7.1 Who Gives a Fuck?

Despite some challenges such as my accident, failed relationships, bankruptcy, and debilitating addiction, my life has generally been blessed with exceptionally good friends and very good advice coming from all sorts of angles.

One of the greatest pieces of advice given was from a very good friend of mine, who provided these words to me in my mid-teens. His words exactly were: 'Who *Gives a Fuck*?'

"Who gives a fuck?" Richard Davies

And that meant so much as, 'who cares what anybody else thinks of you?'

There's a fine line, I know, but, in essence, I think that as long as you're being authentic and true to yourself, and treat those around you with respect, dignity and kindness, then who cares what anyone else thinks of you? And who cares what they say about you to other people, too?

I'm sure that if you stay true to yourself and try to speak only the truth, even just your own truth, then those who matter to you will know that you are being authentic, no matter what anyone else says about you.

---

*"My philosophy is: It's none of my business what people say of
me and think of me. I am what I am, and I do what I do. I expect
nothing and accept everything. And it makes life so much easier"*
Anthony Hopkins

*"Being honest may not get you many friends, but it will get you
the right ones"* John Lennon

*"Confidence is not: 'They will like me'. Confidence is: 'I'll be fine
if they don't'"* Lessons Learned

*"One day it just clicks. You realise what's important and what
isn't. You learn to care less about what other people think about
you and more about what you think of yourself. You realise how
far you've come, and you remember when you thought things
were such a mess that you would never recover. And you smile.
You smile because you are truly proud of yourself and the person
that you've fought to become"* Word Porn

NEAT LIFE HACKS: JUST THINKING ABOUT WHAT OTHERS MIGHT BE SAYING OR
THINKING ABOUT YOU IN A NEGATIVE RESPECT IS NEGATIVE SELF-THINKING,
AND IF YOU'RE NOT CAREFUL, IN THIS WAY CAN BECOME A SELF-FULFILLING
PROPHECY

7.2 OAR

Another great principle I stumbled across, this time during my business
sabbatical, is the acronym, OAR:

O – **Ownership:** take ownership of what you do, and how you live your
life

A – **Accountability**: make yourself accountable for your own actions and behaviour

R – **Responsibility**: take responsibility for the results and ramifications of your actions and behaviour

At a fundamental level it simply explains the importance of taking responsibility for what you have chosen to do, say or feel. It's no good trying to blame anyone or anything for our own actions or behaviour, or the resulting situations or circumstances. Recognise that, for the majority, it's your own behaviour, and take responsibility for that. Only you can change it, if you think that it needs to be improved.

"If you suffer it is because of you, if you feel blissful it is because of you. Nobody else is responsible – only you, and you alone. You are your hell and your heaven, too" Osho

---

There is always something good to be found in every experience, no matter how bad, if you look for it. You might have to dig deep, and some things may seem very bad from the outset, but there is *always good*, and the more you seek it, the better the things will be that happen to you, and around you.

7.3 Always Set Your Intentions

Another neat trick I learned is in setting your intentions before embarking upon doing anything. So, for example, before I go into a meeting that I anticipate may be challenging, I'll tell myself that my intention is to go into the meeting with an open mind, to ask the relevant questions and to listen to all the necessary information without judgement, before making an informed opinion or decision, and to come out of the meeting having made a positive difference or contribution (if it was required), and feeling happy and fulfilled. In the majority of cases, this at the very least sets me up for a good meeting from the outset.

NEAT LIFE HACKS: WHENEVER YOU FEEL CHALLENGED GOING INTO A SITUATION OR ACTIVITY, TAKE THE TIME TO SET YOUR INTENTIONS FIRST, AND FOCUS ON EXPECTING A POSITIVE OUTCOME. INTENTION IS A VERY POWERFUL THOUGHT-PROCESS, WHICH CAN BE USED EFFECTIVELY AT ANY TIME THAT YOU WANT TO HAVE A POSITIVE INFLUENCE ON A FUTURE EVENT. IN FACT, I'D SAY, WHERE POSSIBLE, ALWAYS TRY TO SET YOUR INTENTION BEFORE DOING ANYTHING

7.4 Listen, Think, Speak (LTS) and the Language That You Use

On a similar note, I can't express the importance enough of one of the most significant lessons in communication that I am still working on to this day. And that is the importance of the sequence of listening properly to a person's dialogue/argument, before taking the time to think about it, before constructing an informed, considered and structured answer/comment. I have found this to be so important on a number of different levels, in the relationships that I've built, and the communication that I've engaged in.

The language (or specific word choice) that you use is also of critical importance in life. I cannot stress this enough. I believe that the language that you use sets the platform for the thoughts and emotions that you experience, which are dictated by the beliefs that you store in your deep subconscious, and have a powerful influence on the frequency at which you vibrate. This is something I'm acutely conscious of and am still working on, in terms of how I communicate with you, for example.

Dr Linda Mallory – See Appendix 2

I'd investigated many different methods for working with my core beliefs, most of which were quantifiably successful, such as 'tapping', or EFT (Emotional Freedom Techniques) which I mentioned earlier, or belligerently changing my thoughts as they occur, such as I have also described earlier.

7.5 NLP

Neurolinguistic Programming (NLP) is a very useful source of information when considering the language that you use when communicating with yourself and others, and there is a very simple way of recognising whether the language that you use is raising or lowering your frequency, and that is by observing the way it makes you feel. If the language you use makes you feel bad, then it is lowering your frequency, and if it makes you feel good, then it is raising it. The beauty is that you can choose, or learn to choose the language that you use in order to continually raise your vibrational frequency!

NEAT LIFE HACKS: IF YOU TAKE THE TIME TO ANALYSE THE LANGUAGE THAT YOU USE, RELATIVE TO THE FEELINGS THAT YOU'RE EXPERIENCING, AS A RESULT, YOU CAN BEGIN TO CHANGE THE LANGUAGE THAT YOU USE TO BE MORE POSITIVE, AND SUBSEQUENTLY RAISE YOUR VIBRATIONAL FREQUENCY. IN DOING SO, YOU WILL INEVITABLY FEEL BETTER, MORE POSITIVE AND HAPPY, AND YOU WILL IN TURN ATTRACT MORE POSITIVES INTO YOUR LIFE

7.6 Have No Regrets

In realising that I no longer needed to regret anything, it enabled me to see that in this respect, for as much as I was trying to be a 'good' and 'better' person, I never had to regret anything or feel remorseful about any mistake I made, ever again.

I mean, as long as my intention was never '*bad*', and as long as I learned from the experience, and strove towards a solution, then it had been a *valuable* experience, and I could be happy that I'd had the opportunity to learn. And this became a very powerful tool for much of my life and career.

What I was to learn was that, actually, making mistakes is good, because you can use them to learn, and I learned that, in fact, it is understood that: *if you're not making mistakes, then you're not moving forwards*.

NEAT LIFE HACKS: IF YOU'RE NOT PUTTING YOURSELF OUT THERE AND MAKING MISTAKES, AND LEARNING FROM THEM, THEN INEVITABLY YOU'RE NOT MOVING FORWARD IN LIFE

7.7 When You're Sincerely 'Sorry'

If you've done something wrong, or upset someone, or yourself, then for many people a natural reaction is to be 'sorry', or to want to apologise, and say that you are 'sorry'.

It's my understanding that the true definition of sorry is to accept that you've done something 'wrong', and also, you actively work towards behaviour that will in future ensure that you do not do it again! Otherwise, you can't really argue that you were actually sincerely sorry!

7.8 Asking Stupid Questions, and the Questions that You Ask

Something I learned quite early on during schooling was that I should **never be afraid of asking stupid questions**. In fact, if you have a question, it's *never* stupid (unless it's deliberately stupid, of course). You see, often the question is only as stupid as the information that provoked it. What I mean is that even someone with the best intentions may not have conveyed the information to you in a way that you have understood it. And what are you going to do?

Are you going to spend the rest of the lesson, or conversation, not understanding what they are talking about, and not learning, and being incredibly bored? Or are you going to ask that stupid question, and are you going to put yourself in a better position of understanding, and in the possible situation of learning something new? Trust me, if the person who is talking to you wants to explain something to you properly, then more often than not they will be grateful to you for asking the 'stupidest' of questions.

NEAT LIFE HACKS: DON'T EVER BE AFRAID TO SAY THAT YOU DON'T UNDERSTAND SOMETHING, OR BE AFRAID OF ASKING *'STUPID'* QUESTIONS. IN YOUR QUEST FOR UNDERSTANDING, THERE ARE NO STUPID QUESTIONS, AND, ACTUALLY, SOMETIMES THE MOST STUPID QUESTIONS ARE THE MOST THOUGHT-PROVOKING AND PROFOUND

I've found that the questions that you ask are very important in communication, too. Because, at the lowest level, as described above, it puts the provider in a better position of knowledge on how to more effectively explain something to you (or equally, whether they actually understand it themselves), but at the other extreme, it demonstrates your existing knowledge on the subject.

In my opinion, if you don't understand, and you don't ask questions, then you're not opening your mind to, at the very least, determining whether there is something that you need to, or want to learn.

I chose to view the situation whereby every person who enters, or has entered my life, is, or has been provided to me for me to learn something from. I'm aware that I mustn't judge them or reject them, or assume any authority over them, but must be humble and grateful to them, and understand the lessons to be learned, and to take the appropriate action if necessary.

NEAT LIFE HACKS: RECOGNISE AND ACCEPT THAT EVERYONE ON EARTH IS A UNIQUE AND VALUABLE INDIVIDUAL, AND THAT PREJUDICE AND CLASS STATUS, FOR EXAMPLE, ARE A SOCIALLY IMPOSED POISON OF THE MIND. TREAT EVERYBODY THE SAME, NO MATTER WHO THEY ARE, WHERE THEY'RE FROM, WHAT THEY DO, WHAT THEIR LEVEL OF WISDOM OR INTELLIGENCE IS, OR WHAT COLOUR THEIR SKIN IS, BECAUSE THEY ARE SOULS ON THE SAME JOURNEY AS YOU ARE, JUST AT A DIFFERENT STAGE, AND WITH DIFFERENT PURPOSE

7.9 Sense of Humour

Something else I struggled with throughout my life was my sense of humour. I have always loved to make people laugh and to make them smile and happy. My intention has been to make you laugh a bit throughout this book!

A Theory of Everything

As a child, I developed a method for making people laugh, but, because it was based around *me* making fun of *myself*, and it was predominantly them laughing *at* me, and not *with* me, it was not a healthy method at all, and I needed to change it.

I soon developed a taste for a 'darker' or dry sense of humour, as I began to watch who I saw as being some of the best comedians of this era. For example, Bill Hicks, who used the insanity of social circumstance to point out the negative actuality of our reality in society, and how absurd it is.

"Take all the money that we spend on weapons and defences each year and instead spend it feeding and clothing and educating the poor of the world, which it would pay for many times over, not one human being excluded, and we could explore space together, both inner and outer, forever, in peace" Bill Hicks

"People ask me 'Are you proud to be an American? And I say, 'I don't know, I didn't have a lot to do with it. My parents fucked there, that's about all.' I hate patriotism, I can't stand it. It's a round world last time I checked.'" Bill Hicks

---

At the same time as this began to open my eyes to the global status quo, and the absurdity of our inherent social beliefs and social reality, it sculpted my sense of humour for a long period in my life.

But why do I think that it's important to tell you this? Well, what I realised was, when I became aware of the importance of positive thoughts, language and dialogue, that my sense of humour was based on the negative, and not doing me, or anyone else, any favours. So I had to find a new way of making people laugh, and more importantly preserve my own positivity and level of vibration.

What I eventually realised was that when I was true to myself in my humour, the people who got it the most, and the people who laughed the most, were the people who I resonated with the most, and who I wanted to be around the most.

7.10 Avoid Speaking About Anyone in a Negative Way

I can't stress how important I've found this to be in my life. And put quite simply, when you speak negatively about someone else (or indeed any*thing* else), not only is it likely that you're reflecting your own negativity upon them, you're **allowing** yourself exactly that: to be negative. I've found that this is not only damaging to the person you are talking about, and the person that you are talking *to*, but more importantly, **it is significantly damaging to *yourself*** *as it usually causes a range of negative feelings and emotions, and lowers your vibrational frequency.*

Neat Life Hacks: As a general rule of thumb, I'd recommend that if you don't have anything good to say about anyone or anything, then simply say nothing at all. And once you've recognised that you don't have anything good to say about something or someone, I'd recommend trying to look for something good to think about them or it — no matter how challenging that may be. Equally, take a good look at the negative you have recognised in them, and see if you can identify that trait within yourself

7.11 Complaining

On a similar vein, complaining about your shit isn't going to get you anything other than more shit.

It's the same principle as to why a hypochondriac is always ill, or why somebody who's stuck in a job they don't like, and continually complains about it, but does nothing about it, will inevitably stay stuck in their 'shitty' job.

A Theory of Everything

"See if you can catch yourself complaining, in either speech or thought, about a situation you find yourself in, what other people do or say, your surroundings, your life situation, even the weather. To complain is always non-acceptance of what is. It invariably carries an unconscious negative charge. When you complain, you make yourself into a victim. When you speak out, you are in your power. So change the situation by taking action or by speaking out if necessary or possible; leave the situation or accept it. All else is madness." Eckhart Tolle

---

When you complain about something, you're choosing to express your feelings about it in a negative way. Only *you* can change that. You may not be able to change the situation, so my advice would be to *accept* it, and to begin with changing the way you think about it, to being something positive. I believe that then you are in a better position to change the situation.

For example, Alex, *and I'm looking at you, Alex,* when you find yourself in a position whereby you catch yourself complaining about the fact that you can't have something *right now,* then try changing your thinking to something like, how grateful you are that your 'wanting' is teaching you a lesson in patience, and is giving you the opportunity to think about how you can materialise what you want, or achieve what you want, without hurting or disrupting anyone else.

NEAT LIFE HACKS: IF YOU CATCH YOURSELF COMPLAINING ABOUT SOMETHING, THEN TRY TO CHANGE THE WAY YOU THINK ABOUT THE SITUATION TO BE POSITIVE. IT DOESN'T MATTER WHAT THE POSITIVE THINKING IS, IT JUST NEEDS TO BE POSITIVE, AND I AM SURE THAT IN DOING THIS, IF NOTHING ELSE, IT WILL MAKE YOU FEEL BETTER ABOUT THE SITUATION. AS A BONUS, I BELIEVE, IT WILL ALSO BRING YOU MUCH CLOSER TO GAINING WHAT YOU ACTUALLY WANT

7.12 Don't Beat Yourself Up!

I also learned that there is no point beating yourself up over anything. *Anything* at all! It's far more conducive to progress in your personal development if you *accept* what you've done, or what the situation is, or has been, and concentrate on thinking positively about what you can try to do in the future to prevent it from happening again. Even if it means working extensively with your deep subconscious and subsequent behaviour.

NEAT LIFE HACK: DON'T BEAT YOURSELF UP FOR ANYTHING. WHAT'S DONE IS DONE. THE BEST POSITION YOU CAN PUT YOURSELF IN IS TO ACCEPT THAT YOU'VE MADE A MISTAKE, AND TO USE THE OPPORTUNITY TO ACTIVELY WORK TOWARDS LEARNING FROM IT, AND TO IMPROVE YOUR BEHAVIOUR IN THE FUTURE

7.13 Mixing with the Right People

One of the other things I learned from many sources, including *Unsinkable* and *Think and Grow Rich* was the importance of surrounding yourself with the right people, as I have alluded to before.

It sounds logical, yet if you do a quick reality check and ask yourself if those closest to you bring your energy up or down, the answer may not be what you expected.

NEAT LIFE HACK: REMOVE NEGATIVE PEOPLE FROM YOUR LIFE. NEGATIVITY BREEDS NEGATIVITY AND ONLY ACTS TO ERODE YOUR POSITIVITY AND HIGHER STATE OF RESONANCE (VIBRATIONAL FREQUENCY). YOU CANNOT HELP THESE PEOPLE. TRY TO SURROUND YOURSELF WITH PEOPLE WHO INSPIRE YOU AND BUILD YOUR POSITIVE ENERGY, AND VICE VERSA

Part 8
LOVE AND RELATIONSHIPS
(And a Bit on Sex)

8.1 What is Love?

At some point along the way I'd become confused with the word that we so readily use: *love*. What did it actually mean, or represent? The more I thought about it, the more I realised that, actually, it appeared to be a quite broadly used term that was often rather unspecific.

So, for the sake of clarity, the 'love' that I'm talking about now represents 'good' and 'positive' at its highest level of energy, or light, and apparently with a vibrational frequency of 528 Hz. It is not, for example, referring to the animal instinct that brings two people together, which I assimilate to being akin to a physical addiction (and I have enough experience of that on many fronts to be able to discuss it with some authority!)

"We've been infected with this idea that love is an emotion only felt between two people. But love is universal. An Energy. A contagious force. To offer money to a homeless man is to love. To smile at a stranger is to love. To be grateful, to be hopeful, to be brave, to be forgiving, to be proud, is to love" A. R. Lucas

---

A Theory of Everything

The way I've come to understand it is that people in general tend to hold very different beliefs in their conscious than they do in their subconscious. A lot of the time, the way we think and view life, and how we think we behave, or should behave in our *conscious* mind is very different to the core beliefs (or emotions associated with memories) held in our *subconscious* mind. And because any reaction that happens as a result of our thoughts and emotions is dictated by our *subconscious* beliefs, and *not* our *conscious* beliefs, the behaviour we tend to commonly display, particularly as a reaction to anything happening, is very different to how we consciously *think* we behave, or how we consciously see ourselves. I'm fully aware that this was, and still is to a degree, the case for myself, although I'm far more aware of the difference now.

In that respect, I realised that when you get to know someone, and in particular, on instant chat, for example, or email, or any other form of electronic communication, or when you get to know someone in the first instance of a relationship, *they* may tell *you* what they think that their beliefs and values are from their *conscious* mind and thought, and *you* may tell *them* what you think that your beliefs and values are, from your *conscious* mind and thought. But what transpires is that, actually, you gradually find out what their true core subconscious beliefs are, as you get to know them, and who they really are. It appears people are often unable to see the behaviour that they are actually displaying, relative to how they perceive themselves in their conscious mind. Mainly because they don't seem to understand the relationship that their conscious mind has with their subconscious mind, and the governance that the memories that they've stored in their subconscious mind has on the behaviour that they're actually displaying, despite what their conscious mind is convincing them of.

Human beings often struggle to understand themselves and their behaviour, and how their misunderstanding of themselves affects the ones they love, their communication, and their resulting relationships.

It's my experience that when a couple first get together, they begin to tell each other about themselves from their *conscious* perspective. This is often a very different perspective and persona to that of who they really are at the core, from their *subconscious*. They describe themselves as a person who they consciously believe themselves to be, how they consciously believe that they feel, and how they consciously believe themselves to behave.

They will then often live this persona for a short period of time, if nothing else to prove to themselves that this is who they truly are. It's who, in their conscious mind, they want to be, and see themselves as being. But deep down in their subconscious, they're often very far from being this person, and shortly, when they relax into a more comfortable relationship state, they forget who they were intending to be, and default back to who they really are, from their subconscious memories and beliefs.

The subconscious and conscious beliefs are sometimes two very different scenarios completely. This highlights the importance in this age of texting and online communication that you should really take the time to get to know someone properly, face to face. And that they should really take the time to get to know you properly, too. I think that the best foundation for a long and lasting loving relationship is to first be truly best friends.

NEAT LIFE HACKS: BE AWARE OF THE FACT THAT WHEN YOU FIRST GET INTO A RELATIONSHIP, IT IS LIKELY THAT BOTH YOU AND YOUR PARTNER WILL INITIALLY TELL EACH OTHER ABOUT WHO YOU VIEW YOURSELVES TO BE IN YOUR CONSCIOUS MINDS, AND THAT THIS MAY BE A VERY DIFFERENT REALITY TO WHO YOU ARE AND HOW YOU ACTUALLY BEHAVE, BASED ON THE BELIEFS YOU HOLD IN YOUR SUBCONSCIOUS MINDS. I CONSIDER THAT THE BEST RELATIONSHIPS WORK WHEN YOU ARE TRULY BEST FRIENDS BEFORE YOU ARE LOVERS, OR COMMITTED TO THE RELATIONSHIP

Interestingly, in my previous relationship with Kate, I attempted to address this phenomenon with a structured analysis of my subconscious traits, and a request for her to provide the same. What happened, in combination with a number of other issues, was that my scientific approach to the foundations of our relationship using charts and spreadsheets was not necessarily the best of long-term solutions for this particular relationship!

Here is some very intelligent advice on relationships as presented by Bert Hellinger:

Reference: BERT HELLINGER – Couples Relationship – The secret of love – Part 1 (youtu.be/BOX9o_x-I9E)

Some of the most profound and life-changing strategies and ideologies that I've learned to understand myself, and have accepted with the most remarkable effect, are described below. These have helped me in my understanding of myself and in my relationships with others, and in particular with regard to the stability and harmony in loving relationships:

8.2 The Significance of Loving Yourself First

- Once I was **truly** in a position of **loving myself**, I found that the love that I held for my partner completely changed from what I'd experienced in relationships before. It changed to be significantly more harmonious, natural, sustainable, more easily manageable, more comfortable, and significantly more fulfilling. In this condition, I no longer felt that I *needed* a partner to satisfy my desires or needs, and that I no longer felt any *dependency* on my partner. I no longer felt *deluded*. The love I felt was more pure and unconditional than I had ever experienced before, and it was truly amazing. As importantly, I found that the love that I was then subsequently able to give, was received and magnified in a way that I had equally not experienced before

8.3 The Significance of Having No Expectations and No Demands

- Once I accepted the philosophy, which I originally came to understand as quoted by Anthony Hopkins, and put myself in a position whereby I had **no expectations and no demands**, then I truly became at one with myself, and my loving relationship became significantly easier for me to manage emotionally, and significantly more harmonious. I found that this sentiment was exceptionally powerful when addressing my sexual desires, previously associated confusion, and lack of confidence. Yes, it has a lot to do with how amazing, grounded, stable, and broad-minded my partner is, too, but it enabled me to fit better into her world, and to integrate more synchronously with her

8.4 The Significance of Accepting Your Partner for Who They Truly Are

- Acutely conscious of the fact that I couldn't change anyone, once I truly understood the value behind Bert Hellinger's message quoted above, and allowed myself to **accept and embrace every behavioural trait in my partner** as being an important part of her, then the relationship became significantly easier and more harmonious for us both. Not only that, but this in turn put us both in a position to more easily and naturally observe the behaviours in each other that upset the other, and to understand each other's negative reactions in the short term, as simply coming from deeply ingrained subconscious memories from the past

8.5 The Significance of Possession Being the Opposite of Love

- To quote a phrase used by one of the Gods in 'Vikings', and put quite simply, '**possession is the opposite of love**'. It would be easy to equate this to being an excuse for him to justify his shagging every attractive young woman in the settlement; however, for me this was an exceptionally powerful sentiment, and enabled me to dissolve any jealousy or insecurity that I may have otherwise harboured, and to truly value and trust the love that was being endowed upon me, absolutely unconditionally. Some time ago, I realised that I did not ever want to assume any governance or power over anybody, and this put the icing on the cake for me, and it has made my life and love far more satisfying and fruitful

8.6 The Significance of Recognising a Mirror in Your Partner's Behaviour

- Once you realise that often you see your own bad behaviour as a mirror in how you perceive your partner to be behaving, you can begin to question whether it is actually *you* who has the issue that you are unconsciously blaming your partner for.

When we observe character defects in other people and criticise them, it's really the undeveloped parts of our personality that are showing up. We're only irritated by these blemishes because the very same issues are unresolved within ourselves. When we see faults in others it can be used as an opportunity for self-reflection.

If we think someone is arrogant, we can examine our own egos. If we describe someone as being unkind, we can examine our level of kindness, compassion and empathy. If our friend's judgemental nature bothers us, we should think about how we view other people.

We should always endeavour to look at people in a positive light. But when it becomes difficult, it is an opportunity to look inwards.

NEAT LIFE HACKS: TO PUT YOURSELF IN A BETTER POSITION TO GIVE AND RECEIVE LOVE UNCONDITIONALLY IN A RELATIONSHIP, YOU MUST FIRST TRULY LOVE YOURSELF. YOU MUST PUT YOURSELF IN A POSITION WHEREBY YOU HAVE NO EXPECTATIONS OR DEMANDS. YOU MUST ACCEPT AND MAKE ALLOWANCE FOR WHO YOUR PARTNER TRULY IS, AND YOU MUST HOLD NO GOVERNANCE OVER THEM

8.7 Sex, Sex, Sex

Yes! **SEX!**

Well, there's an interesting one!

In a previous chapter, I spoke about us not being born with a guidebook for life. Therefore, depending on a person's religious belief and cultural background, sex may be viewed differently. In the UK, for example, we mustn't talk about it, but most people, deep down want to talk about it, if we allowed ourselves to do so! Some are desperate to talk about it, but generally speaking we don't, and certainly not with much sincerity. And we hide it from you, our children, and we make it allusive and secret (note: and sacred). So much so that when you get to the right age, you may even develop a pornography-fuelled obsession for it, that which has been withheld from you. It becomes unhealthy and unnatural, and erodes all of the natural magic and wonder associated with it.

Where does love fit into sex? I don't think that many people understand this question very well, and behave as if they think they should behave. I don't blame them as it is 'normal' and it's as they've been told and conditioned to think they should behave, and exactly how I used to behave. So-called modern societies around the world appear to have conditioned us to suppress sex, and have driven it underground to seedy arenas and backstreet whorehouses. They've turned it into an industry! Surely nobody should live in a society where they are so desperate for sex that they would want to, or have to pay for it; or conversely, sell their bodies in this way?

"We live in a world where we have to hide to make love, while violence is practised in broad daylight" John Lennon

---

A Theory of Everything

For many, sex is for those who embrace the notion of 'eros' (for example, *the sum of life-preserving instincts that are manifested as impulses to gratify basic needs, as sublimated impulses, and as impulses to protect and preserve the body and mind*) defined within a committed and monogamous relationship. It's for those with the wisdom to understand the implication of it, and its deepest meaning as an expression of profound love and dedication towards each other in such a relationship. Here, it's a union of spirit, soul, and body, to those who are deeply 'in love'. In some cultures, sex must be consummated only in the context of marriage or commitment vows, where other cultures are more open in their approach and beliefs.

Allow me to explain this topic a bit more. In some cultures, sex is regarded as merely a biological need, and regardless of any relationship a person is in, they indulge in sex with other people. They call it free sex. I'm not a moralist. I'd rather be more understanding, and respectful of all cultures and their convictions, and without judgement towards anyone.

For those who believe that sex is merely a biological need, I choose not to blame them as our sex drive is 'human'. My advice to you in our current environment is to initially practise 'safe' sex, because for whatever reason, there is a chance that sex with a stranger might be detrimental to your health, as some diseases can be contracted. More importantly, an unwanted pregnancy can cause major ramifications in your life!

Whatever religious or cultural belief you have, something I'd like to impart to you is that sex doesn't have to be, and is not 'dirty' at all. Sex can be sweet and tender and pleasurable, and if orchestrated and conducted in the right way, and in the right context (privately and safely) and with mutual and informed consent, it is one of the greatest pleasures on earth. In fact, you may even make the earth move (quote from *The Sun Also Rises*, Ernest Hemingway)!

Book Reference: *Fiesta. [The Sun Also Rises].* Ernest Hemingway. (tinyurl. com/2p8hdudu)

"When the sun rises, it rises for everyone" Aldous Huxley

8.7.1 ORGASM

The orgasm is widely known as the climax of sexual excitement. It is a powerful feeling of physical, and spiritual pleasure and sensation. It's understood that the point of orgasm is the point at which one can instantly, if momentarily, become at one with oneself, in the absence of any ego or any external or internal stresses, and that this state can otherwise be reached through meditation.

It's also understood that if you guys, and children around the world, were introduced to meditation at an early age, and if we, your parents and society, were more open with you surrounding sex, whilst at the same time protecting you until you are old enough to engage, not only would you develop a far more healthy understanding of sex, you would also not need protecting from it, because you would be in a better position to accept it for what it really is in its ultimate power and beauty.

But more concerning is the explanation of why 'they' have done it. 'They' have disassociated sex with the full magnitude and power of the infinite possibility of the positive frequency of love, and the significant and life-changing spiritual benefits associated with it. For it is understood that if you truly knew the significance of the point of orgasm, which deep down we all crave, then you might understand life and your existence itself! And 'they' wouldn't want that at all, because it would give love and the human race ultimate power over them. As I said before, I'm convinced that if us 'normal' human beings got together and united in unconditional love, then the magnified frequency of Love would kick Evil's arse!!!

Just think about it. If what I'm saying is true, imagine the scale of the organisation and proliferation of the lie that has had to be installed globally over millennia, through multiple channels, in order to control ~ 7 billion people in this way! So, either you think that I'm crazy and talking shit, or alternatively, you could choose to open your eyes to the magnitude and significant influence that 'they' have managed to instil into our societies and cultures at every possible level, and you could unite in love to combat the evil that is coming to its climax in this era!

Some of what I've described above is reflected in one of Osho's many books (which, I understand, are not for everyone), *From Sex to Superconsciousness*, where he appears to conclude that the ultimate goal of understanding the point of orgasm is that of ultimate celibacy. This is interesting, since, understanding a little of his history, he appears to have spent most of his life having intercourse with as many women as physically possible, and had no regard for the children that he left behind. Oh, and he started a controversial cult that received much negative press. But that aside, I think that much of what has been written about what he has taught (because, I guess, he was so busy having sex that he never bothered to learn to write himself) makes a tremendous amount of sense to me in the bigger picture, and in particular with regard to my understanding of the soul, and finding our true success.

Osho's authors have provided much more detail on his understanding of how 'they' have orchestrated the global social situation over the millennia, and I'd recommend reading at least one of his many publications for an insight, because, as I said, much of it makes a lot of sense to me, with respect to the bigger picture:

A Theory of Everything

Book Reference: From Sex to Super Consciousness by Osho (tinyurl. com/mw6jzut5)

I have not read them myself, but would suggest that if you want, you look at the following more specific books from him relative to Children, Men or Women:

Book Reference: *Book of Children* – Osho (tinyurl.com/5n9xw2c5)
Book Reference: *Book of Women* – Osho (tinyurl.com/2p8uvn4m)
Book Reference: *Book of Man* – Osho (tinyurl.com/h5rx73ze)

"Lightning shines in the darkness of the night, but the darkness is not part of the lightning. The only relation between the two is that lightning only stands out at night, only in the darkness. And the same is true of sex. There is a realization, an exhilaration, a light that shines in sex, but that phenomenon is not from sex itself. Although it is associated with it, it is just a by-product. The light that shines in orgasm transcends sex; it comes from beyond. If we can comprehend this experience of the beyond we can rise above sex. Otherwise, we will never be able to." Osho

---

For me, the truth about sex is an exceptionally important topic in education. An education in the absence of malice, and where proper provision of guidance is available, especially from an early age. For example, this would appear to be available to children in Holland, Europe. So, don't be afraid to ask questions of your parents, elders, teachers, and religious leaders, and read about sex. Don't get embarrassed if you're caught talking openly and sincerely about sex. Sex is a part of your existence. It is a gift to you, so use it appropriately on your path to divine ascension!

Have fun, be safe, and be responsible and sacred!

Part 9
My Healing and Wellbeing
Principles – Summary

9.1 Nutrition, Medicine and Healing

Wow, this is a big topic. And one to which I have dedicated much of my research and time. Mainly from my desire to heal my own physical pain and ailments when conventional medication failed to be effective for me.

So, I'm just going to come out with it and say it. In my years of research, I've learned that:

"The pharmaceutical industry in general, and at the very highest level, is a very powerful corporate-run body who do not care about anybody's health, wellbeing or recovery. They only care about making money and selling you their chemically synthesised drugs and treatments. Some have developed these drugs and fabricated the data and the education system for the primary purpose of not making you better, but prolonging your illness in order to sell you more drugs, and make more money. Generally, and again, at the highest level, they do not care how they do it. It is at least a multi-billion-dollar industry, and many of the common ailments known to man are rumoured to have been equally synthesised by these exact bodies for the purpose of making more money from the general population" Dr Jonathan R Tuck

And, in terms of healthy eating and nutrition, I will also say this:

"The food industry in general, and at the very highest level, is a very powerful corporate-run body controlled primarily by 8 of the top global corporations, including pharmaceutical companies, who do not care about anybody's nutritional health or wellbeing. They only care about making money from you and selling you the food that they have produced in the cheapest way possible, regardless of health or nutritional value. It is at least a multi-billion-dollar industry, and they really don't care about us, or the food that we eat or our resulting health" Dr Jonathan R Tuck

---

You can quote me on all of the above. I don't care. That's the way I see it. Many people understand this to be the case, and if you look at the evidence and do your research, then you will see them all to be absolute fact, and, as I said, well understood by many people.

That said, I have the utmost respect for many health practitioners, especially those who saved my life after my childhood accident, and also many health professionals and business people working in the health and nutrition industry, who are sincerely dedicating themselves to the wellbeing of humanity. They are unequivocally excluded from the above tirade!

9.2 General Health and Wellbeing Checkpoints

As I understand it, there are several key parameters that are fundamental to human health and wellbeing in general, based on the observation that the body is an electrochemical system that predominantly consists of water:

- **The body's pH level**
  - o No detrimental bacteria or pathogen can exist in an alkaline environment
  - o Please see section 9.7.2 for information on how to achieve and maintain an alkaline environment in your body
- **The body's voltage**
  - o Electromagnetic radiation and environmental toxins all contribute to raising your body's voltage, and putting things out of balance and harmony
  - o Please see 9.14 and 9.15 for information on how to replenish electrons in your body, and to bring your voltage back into harmony
  - o Prior to contrary belief, salt is very good for you as it forms an electrolyte with water to help the body work. I use Himalayan Sea Salt
- **The body's natural vibrational frequency**
  - o The body is more healthy and vibrant when your cells are vibrating at a higher, more positive frequency
  - o Please see 5.6 for information on how to raise your vibrational frequency

When people get ill it's usually through either physical accident, nutrient deficiency, or toxic poisoning.

Generally speaking, you will die of either an 'itis' or an 'osis' (clivedecarle. com):

  o itis – inflammatory disease
  o osis – scar tissue – sulphur dissolves scar tissue (see 9.5.5)

9.3 That Gut Feeling

As I mention at the beginning of this book, we are born with a very powerful guidance tool that is often overlooked, and that is our **intuition,** or our **gut feeling**.

The term 'gut feeling' is interesting. And there's no coincidence either.

The feelings or sensations we often experience in our gut that tell us whether something feels right or wrong appear to stem from something known as the 'enteric nervous system' located in the stomach area. This part of our nervous system is also called our 'second brain'. It's understood that around 80% of key neurotransmitters for mental health are manufactured here. Neurotransmitters are chemical messengers that impact our mood, appetite, drive, etc.

Both our main brain (the one I tossed about as a child) and our second brain (our third brain is the heart) and the vast network of nerves that connect them play key roles in regulating our emotions and thought processes. They greatly influence how we interpret information and make decisions. Our three brains are in constant communication with each other, conveying information observed from your sensory receptors (sight, touch, smell, hearing, etc.) for processing, enabling you to initiate a response. Because our second brain, the proposed source of our intuition, is located in the digestive tract, it is now understood that many of our mental and cognitive imbalances are also in part related to the food that we eat and the general state of our digestive tract. What's more, the trillions of bacteria found in our digestive tract also play a crucial role in the way we think, act and feel. The link below is a good resource expounding the scientific evidence behind this phenomenon:

Reference: Evidence mounts that gut bacteria can influence mood, prevent depression (tinyurl.com/4uehm7w3)

In short, your gut and its microbe inhabitants have a direct influence on your mental state and how your brain processes information. A healthy gut makes an excellent foundation for a healthy mind. This means that the healthier you eat, the better your mind will work and the more balanced your emotions are. I therefor advise you and anyone strongly to ditch sugar and all processed foods as soon as you can if you want your brain(s) to function well!

Don't take antibiotics unless strictly necessary. Antibiotics have the potential to kill many of the valuable bacteria that keep out gut healthy.

Documentary Reference: Your Amazing Body (AB) (tinyurl. com/2p9d5233)

Book reference: *Gut: the inside story of our body's most under-rated organ: new revised and expanded edition* by Giulia Enders

9.4 Toxins

There are over 100,000 recognised toxins that result from modern consumables in every arena, including the food that we eat, the medicines that are prescribed to us, the vaccinations that we've been treated with, the by-products of industry and machinery, pesticides, herbicides, the cleaning products that we use, the cars, boats and planes that we get around in, the cosmetics and toiletries that we use, and even the water from most plastic bottles that we drink. The list is extensive. These toxins clearly do not kill people quickly; however, they are worth identifying and removing periodically, and where possible, avoiding.

One good example of avoiding something toxic which is regarded beneficial: Commercial sunblock. Avoid using commercially available sunblock. I learned that from my academic peers during my time at the University of Oxford, so I trust the source. Most of these products are loaded with chemicals and actually create free-radicals in the body. They can cause cancer, rather than prevent it. If you're going to use sun cream, then opt for an organic, mostly chemical-free version. I'd recommend the following company:

Reference: Green People (tinyurl.com/2pkjvb5p)

Equally, where possible, I'd recommend avoiding drinking water out of plastic bottles, particularly those containing Bisphenol A (BPA). This is one of the most common sources of toxins that we are exposed to on a routine basis. If you can't avoid it, because this is very challenging, then at least ensure that you minimise the exposure of the plastic bottle to direct sunlight, and when you drink from them, you avoid creating a vacuum when you suck on them, as this causes the toxins to leach, or be pulled into the water. Here is a link to some information on the topic, and a list of some water bottles that do not contain the chemical:

Reference: List of BPA free bottled water brands (tinyurl.com/2p9ehrux)

9.4.1 DETOX

There are many ways in which you can detoxify your body and remove unwanted metals and chemicals, and I'm going to describe the ones that I have direct experience of. These consist of 'plant medicine', pulsed electromagnetic frequency healing, herbal detox, and dietary detox.

9.4.1.1 KAMBO DETOX

The most powerful and immediately tangible detox substance that I've experienced is a remedy called Kambo, which is an extract from an

Amazon Rainforest frog. It **must** be administered correctly under the supervision of a **trained professional**, because it's essentially a poison, and can be very detrimental to your health if taken in doses that are too high. When I first experienced the effect of Kambo, I regained energy that I had not had for a decade previously, and almost immediately. It's not for everyone as it's an albeit short-lived, but unpleasant experience, where it induces vomiting after about 20 minutes of application. However, you can see the results of the detoxification immediately in the form of the blackness of the toxic bile that is being expunged!

Reference: The Ultimate Guide to Kambo (tinyurl.com/2nncvdb3)
Reference: Psychedelic Times – Kambo (tinyurl.com/ch7bmj8x)

9.4.1.2 PULSED ELECTROMAGNETIC FREQUENCY DETOX

A more harmonious method for detoxing heavy metals and contaminants from the body is by using remote frequencies such as applied by Rife, or Spookey2 equipment. Here you can, for example, choose programs that remove either with or without mercury, depending upon whether you have any amalgam fillings. This is because if you have any amalgam fillings, and you remove mercury from your body, then your fillings will become loose and fall out, which is counterproductive and detrimental to your health!

9.4.1.3 HERBAL DETOX

I have previously worked with Alicia Sawaya (aliciasawaya.com) who is a spiritual herbalist, who used herbal teas and complex routines and preparation schedules to detox my liver and kidneys.

9.4.1.4 DIETARY DETOX

There are many detox diets out there if you go looking for them, and there are so many that I am not going to reference any here. I'll just say that most of the detoxes that I have used have been very beneficial to my health across the board, and have also contributed to my weight loss. Many focus on reducing the amount of food that you consume over a period of time, and also improving the quality of the food that you eat, and use specific food groups for specific purposes.

Fasting is a very good way of detoxing, and several times I have performed a detox where I consume nothing more than water and black tea (and I don't care what anyone says, black tea tastes terrible!) for four nights and three days. I found that the first day was relatively easy. The second day was challenging; however, on the third day the sun came out and I experienced great mental clarity and energy, and the tangible effects of the detox.

9.5 Nutrition Insights

9.5.1 EAT HALF (HENNING UHLENHUT)

What I've learned is that as human beings, particularly as adults, we really don't need to eat as much as we're led to believe, and that is being sold to us, and literally shoved down our throats.

For your daily dietary needs, depending on your health condition, you may consider consulting a dietician or nutritionist.

We live in a society of abundance, and if you're not careful, after the age of about 30, that abundance will make you fat and unhealthy (the 'middle-aged spread'). Trust me, I know!

Also, my initial response to the advice I was given by Henning is not advised. When he told me to 'eat half', I went out and bought *two* chocolate bars!

9.5.2 AVOID EATING BREAD, OR ANY (WHITE) FLOUR PRODUCTS

Whole wheat gluten-free bread and unleavened bread are more favourable, especially if the wheat was grown organically. Bleached white flour can contain bromine, which prevents iodine absorption which can affect thyroid function. It can contain alloxan, which can contribute to developing diabetes.

It's postulated that white flour was developed (in the UK) during WWII as a method for making bread with a normal shelf life of 3 days, to have a shelf life of 3 weeks. Well, they did this very successfully. However, what they created was a 'flour' produced from wheat grain that had been completely stripped of its goodness, and as a result, when ingested, creates a massive spike of insulin in the body, which soon crashes, and is very disruptive for your blood sugar levels and general wellbeing.

Modern wheat flour also contains higher levels of gluten and pesticides which are giving many people all sort of digestive issues.

9.5.3 AVOID EATING REFINED SUGAR

Yes, you heard me correctly. **AVOID EATING REFINED SUGAR.** *And I'm looking at you, Alex!*

I believe that sweets and sugar (and particularly refined sugar) are incrementally very bad for you, and should be avoided at all costs, particularly in combination with fatty processed products, including crisps, biscuits, commercial chocolate, and other cheap snacks or ready meals.

Consistent consumption of refined sugar can lead to gut problems, reduced brain function, obesity, diabetes, heart disease, and a range of other inflammatory diseases later in life including cancer.

If you're going to eat chocolate, I'd recommend eating 90%+ cocoa chocolate, or high content raw cocoa chocolate that has been responsibly sourced. Or you could make your own with raw cacao powder which isn't heated or processed and considered a super food! This has good antioxidant properties, and for me is a treat that I can be happy about eating in moderation.

To replace refined sugar that you add to your diet yourself, consider coconut sugar or Stevia, which is made from extracts from the leaves of the stevia rebaudiana plant.

9.5.4 SUPPLEMENT YOUR DIET

Supplement your diet, for example with organic unfiltered apple cider vinegar and sprouted seeds (Chris New):

The enzymes in sprouted seeds are great at helping to break down the food you eat. The foods highest in live enzymes are sprouted seeds and pulses. There are about 10 to 100 times more live enzymes in sprouted seeds and pulses than there are in most vegetables or fruits. A great digestive booster.

- You can buy these from health shops or even some supermarkets, but if you want to make your own:
- Put some seeds / pulses into water. I use alfalfa seeds, or mung beans, but you can use other vegetable seeds, like radish or broccoli or even chickpeas
- Allow them to soak for about 24 hours
- Rinse them thoroughly in clean water
- Repeat this process until they have begun to sprout (1-3 days), and

then rinse them and refrigerate them for consumption with your meals

9.5.5 VITAMINS AND MINERALS

There are many supplements you can take in this category, so I am going to cover the essential components that are commonly found to be lacking in many people, which when administered can help with a tremendous amount of healing at many levels. Minerals do not decay; however, it is important to boost your level of vitamin intake because vitamins only survive in vegetables for a small amount of time after harvesting, so even if you think you're eating the appropriate quantity of vegetables, if you're unaware of the source, then there's a chance that the level could be anywhere to zero in the food that you're eating.

- Vitamins:
  o Vitamin C
    - One of the most powerful vitamins for healing and wellbeing is vitamin C. It's not only good for general day-to-day wellbeing, but it's also very effective in clearing bacterial infections and contributes to boosting your immune system
    - Be careful what you buy and where you get it from. It's understood that approximately 98% of this vitamin on the market is made from GMO maize from China, and is not so beneficial to your health, so do research your source appropriately
    - The best delivery system you can buy is in a liposomal form. When we take the regular tablet or salt form, we only tend to absorb 20% of the vitamin, whereas with liposomal form about 80% is absorbed, so endeavour to get liposomal form where you can

- The only effect of 'overdosing' with this vitamin is resulting diarrhoea, which can be used as a measure of it having completed its task, if administered correctly and for the right purpose
- The advice is to take 2000mg every 15 minutes until you have a bowel movement, after which point, you'll know whether it's been effective or not
  o 'Vitamin' D3
    - Vitamin D is not actually a vitamin, it's considered to be a hormone, and as you're probably aware, can be readily absorbed by exposure to the sun, and particularly through the eyes in a practice referred to as 'sungazing' (see 9.8)
    - Again, one of the most effective forms is liposomal
  o Vitamin B12 and K

- Minerals
  o 1. Magnesium
    - Magnesium oxide is good for constipation, and also dissolves calcium
    - Too much orally will give you diarrhoea, similarly to vitamin C; however, rubbing it into the skin does not give this problem, or bathing in it – look for Epsom bath salts
    - It's good for pain relief and stress relief
    - You can take magnesium capsules and the advice is to take up to 12 capsules depending upon how much you need to treat, or how bad you feel
    - Also see magnesium citrate (magnesium chloride)
  o 2. Iodine
    - This assists with memory and temperature regulation, and for dry skin

- Iodine also controls the thyroid – see Lugol's iodine (link to resource below)
- Take 12.5mg in water a day and then 50mg for 5 days a week for 6-8 weeks for addressing memory issues

Book Reference: *Iodine: Why You Need It. Why You Can't Live Without It* – David Brownstein (tinyurl.com/5t4sp8kt)

- o 3. Selenium
  - It is best to take selenium alongside iodine
  - Selenium is found in Brazil nuts
  - Also see selenomethionine (amino acid)
- o Soil
  - Soil is one of the most powerful mineral substances and detoxifying agents
  - The source needs to be virgin and untouched by modern agriculture and contaminants such as fertilizers and weed killers, etc. There are several places in the world where the soil is very nutrient rich
  - Put the soil in some clean water and consume
  - If you're not comfortable consuming Baldrick's substitute coffee (Blackadder 4 – UK comedy from the 80s) then you can buy fulvic minerals, or fulvic acid. Take a few drops a day in pure water with no chlorine or fluoride
- o Sulphur (dissolves scar tissue)
- o Boron (boost testosterone and helps with arthritis)

Much of this information is taken from Clive de Carle who has been an invaluable mentor and teacher to me surrounding the truth behind health and wellbeing, and how to cure dis-ease: clivedecarle.com

9.5.6 EAT A MOSTLY MEAT-FREE, NATURALLY COLOURFUL, PLANT-BASED DIET

Where possible, attempt to eat small quantities of meat (and fat), and focus your intake on an organic plant-based diet, which has been grown in the absence of pesticides and chemical stimulants. Where possible, it's advisable to grow your own food, so that you can be sure of its source and purity.

It is understood that low fat diets can damage the brain, since the brain is predominantly fat and water. That's why Alzheimer's was first diagnosed shortly after margarines and processed foods came into circulation. This can be treated with coconut oil, which reduces high cholesterol, and can be performed over a 3-month course: clivedecarle.com

9.5.7 AVOID DAIRY PRODUCTS

For example, contrary to what we've bought into, milk is for calves, and not designed for humans! I'm equally conscious of the fact that commercially mass-produced milk is homogenized and the 'goodness' is predominantly destroyed in the process. Also, there is a school of thought that suggests that these animals are likely to be kept in poor living conditions and treated badly, and that this negative trauma and experience is passed to us through the milk that we obtain and consume from them. I can very much believe it, with my understanding of Universal Energy. In fact, for this reason I'd say that if you're going to eat meat, then you'd be very wise to consider eating truly organically raised animals that have experienced a happy and harmonious life, while they have been on this planet.

9.5.8 USE AN ARRAY OF HERBS AND SPICES

Herbs and spices all have their own healing benefits to the body, and consider consulting a spiritual herbalist, such as Alicia Sawaya (aliciasawaya.com).

9.5.9 TWO FRUITS TO CONSIDER EATING DAILY

- Organic Apples – but don't just eat the flesh; it is understood that the seeds are very good for you, and it is even suggested that they can help to address cancer and other ailments
- Lemons – ½ an organic un-waxed lemon a day will generally help your body to function more optimally – to 'un-wax' a waxed lemon, simply pour boiling water over it until it's clean – I usually put the lemon into a mug and pour boiling water over it twice until the residue is completely removed

9.5.10 LOOK AFTER YOUR TEETH

Do not accept amalgam fillings in your adult teeth. Hopefully, you won't have any fillings. But if you do, then always insist on having a composite filling

- The body is an electrochemical organism, and it is thought that each one of your teeth is responsible for the health of one of the other organs in your body
- **Book Reference:** Healing is Voltage by Jerry L Tennant
- Consider the practice of 'oil-pulling' (your mother used to do this) to keep your teeth clean
  o Reference: Coconut Oil Pulling Benefits and How to Do Oil Pulling (tinyurl.com/4xd27j8f)

9.5.11 DON'T DRINK ALCOHOL EXCESSIVELY

- Dudes, you will probably go through a stage in your life whereby this 'drug' has some influence over you or your environment and peers. And who am I to preach, based on my history, and the experiences that I've had?
- The only thing I will say is to bear in mind that it is the fact that **alcohol is a poison**, which, when consumed in excess, or on a regular basis, **will kill you slowly**. If you ever have any issues with it at all, please talk to me. I will not judge you and I will understand, and I will endeavour to provide you with the best advice and support that I can

9.5.12 LEARN TO FORAGE

- Did you know that there is an absolute abundance of edible plants pretty much right outside your doorstep? Most people call them 'weeds'. I call them golden veggies. Dandelion, clover, daisies, plantain, nettle, and many more. They are all edible and all exceptionally high in nutrients. I reckon just two handfuls of home-picked 'weeds' (in a non-pesticide sprayed area!) possibly contain far more vitamins and minerals than an average pre-packed bag of salad from your local supermarket. It's well worth learning what you can and can't eat in nature. Not only is it healthy, it's fun, too! As well as totally for free.

Book Reference: Food For Free (Collins Gem) Paperback, 12 Apr 2012, by Richard Mabey

9.6 CBD Oil

The two main active ingredients of cannabis are THC (Tetrahydrocannabinol) and CBD (Cannabidiol). THC is the psychoactive component. Balancing the ratio of THC to CBD in the plant is the key to its special and diverse properties. However, it has been known for a long time that the CBD component is extremely beneficial to health, and this product is now legal in the UK.

Reference: What Are THC & CBD? | Marijuana (tinyurl.com/35f3nf2h)

You can buy this product from reputable health food shops, and the additional advice provided is to follow the dosage instructions carefully, and when you take it, to hold it under your tongue for two minutes. There are different strengths available at varying costs. Depending on the condition you use it for. It may be a bit of trial and error on dosage and strength, but rather opt for quality than quantity!

Reference: Love CBD dosage instructions (I'm not advocating this brand, purely using it for the dosage reference) (tinyurl.com/yhffmrry)

I am currently experimenting with this for pain relief, but I have noticed a number of other interesting and beneficial side effects, such as better sleep resulting in less fatigue, which in turn results in better pain management.

9.6.1 WEED

Yes, I said Weed!
Cannabis, Marijuana, Dope, Solids, Pot, Ganja, Hash, Herb, Chronic, Reefer, Mary Jane... whatever you want to call it, in its purest and most

natural, and most appropriately balanced form (relative to what you're trying to treat or address), it is one of the most powerful and beneficial plants on the planet, for both body and mind. **End of story**. If you understand it to be anything other than this, then let me tell you, you have been lied to!

You see, historically, mankind, in many different civilisations and cultures, has used this plant for millennia and before, to treat illness and pain, and for working with the mind. However, many years ago, when 'big pharma' realised that they'd make far more money if this naturally available plant were *not* available to the general public, they started the process of closing it down as much as they could. Like with sex, they poisoned our minds, and drove it underground. Interestingly, hemp was also outlawed around about the beginning of the 20th century, but that's another story.

Yes, people can abuse cannabis, in a similar way that they can abuse alcohol or any other substance, but it's understood that not a single person has actually died from administering it, whilst not also under the influence of anything else at the same time. Equally, you have to be very careful of the source and purity of the plant that you're buying. If any of the horror stories that you've heard are true, then they're very likely to have emanated from somebody using something impure or contaminated, for whatever reason. If in doubt, don't use it and always opt for quality over quantity.

This plant, when pure and administered wisely and appropriately, will introduce you to your mind, initially at its deepest and darkest levels, but with balance and experience, will show you who you are, and what you *really* want to do with your life, and the choices that you have. It will introduce you to your*self*. It will help you to work with your subconscious, if you'll allow it and work with it, and it'll provide you with the most powerful clarity of thought, once you've learned how to use it, and once you've begun the process of mastering your mind and thoughts. It'll provide you with direct access to your intuition, and unlock the genius within you, which I believe is within many of us.

"You're beginning to understand, aren't you? That the whole world is inside you: in your perspectives and in your heart. That to be able to find peace, you must be at peace with yourself first; and to truly enjoy life, you must enjoy who you are; and once you learn how to master this, you will be protected from everything that makes you feel like you can't go on, that with this gift of recognising yourself, even when you are alone, you will never be lonely." The Spiritual Compass

---

It can also help with, if not cure, many ailments and conditions, including assisting with sleep and pain relief. It's helped me tremendously in both respects (mentally and physically), and more profoundly and immediately so, than anything I've ever had the pleasure of experiencing before.

There are the two primary components of cannabis, which enable this profound phenomenon:

- CBD (Cannabidiol)
  o This is essentially the healing component

- THC (Tetrahydrocannabinol)
  o This is the psychedelic component

This is quite a good link for some information, but if you do some searching around, you'll find a plethora of non-biased information on the different strains of this plant, and its uses and benefits:

Reference: THC, THCA, CBD, CBC, CBN: Medical Marijuana Composition, The Chemicals in Cannabis (tinyurl.com/2p829jyv)

It's more complicated than I've explained above, for example, there are different types of each component, and the art is in being able to match the type, and ratio of CBD and THC, to the specific purpose.

Having said all of that, I wouldn't necessarily recommend that at your age, having read this, you should immediately go out and score some weed, skin up and get stoned. Don't forget that your 'brains' are still developing till the age of 21, so if nothing else, be careful what you're doing, and if in doubt, please ask me for some advice! At your age I'd recommend meditation as a safer tool, before considering weed.

9.7 The Water That We Drink

9.7.1 THE QUALITY OF OUR WATER

Depending upon the geographical area that you live in, and certainly for many of the States in the US, the water that is delivered to our taps is for a large part deliberately contaminated with fluoride. It's presented to the general population as being for the purpose of preventing tooth decay.

Bull... Shit! Is all I can say. I mean, really?! And most people still believe this... They've done a fantastic job at making us ignorant and proliferating the lies. Even many dentists have been brainwashed into believing it (or have been bought into it).

The primary reason fluoride is added to the water is to gradually dumb down the general population without them knowing it or realising it, and keeping them as separate from their spiritual side as possible. The same reason that alcohol is legal (it's called a spirit for a reason).

In some developing or third world countries tap water is contaminated with chlorine and other chemicals that are not good for the human body. Drinking chlorine with your tap water is known to cause some adverse health effects. These effects occur because chlorine tends to form trihalomethane (THMs), including chloroform. These chemical compounds have been found to cause adverse health outcomes. Below are some of the potentially harmful effects of chlorine water: asthma symptoms, food allergies, congenital abnormalities, bladder and rectal cancer.

Reference: Long's Eco Water System (tinyurl.com/yvjn9svf)

The best water that we can drink is spring water from a mountain source. Throughout history, humans have mostly drunk untreated fresh water apparently until about the last 100 years. Fresh healthy drinking water has existed almost everywhere in nature, especially around mountains or valleys. Water from mountain springs often contains a considerable amount of minerals. Water with high mineral content such as calcium, iron, magnesium and sodium is beneficial to our health in the long term. It could also build up antibodies to bacteria and other impure substances over time. However, with forest denudation, and soil pollution, we now depend heavily on treated water.

9.7.2 TREATED WATER FOR BETTER HEALTH BENEFITS

- Drink 'alkaline water' daily and try to drink only filtered water
- The body's pH level
  - This should be around pH 7 and is affected by the toxins in our environment, the food and drink we consume and our stress levels (Please meditate regularly!)

o Most pathogens, or harmful bacteria can't survive in an alkaline environment and inflammation is less in an alkaline environment
o Drinking alkaline water regularly in the morning on an empty stomach can lower your pH level to a state where you can be much healthier and less susceptible to disease (please see below for details)

9.7.2.1 HOW TO MAKE ALKALINE WATER

- Slice 1 x non-waxed lemon, preferably organically sourced
  o If you cannot get an unwaxed lemon, then simply wash the lemon in boiling water to remove any residue or contamination before cutting it
- Put into 2L **glass** vessel and cover with 1 tablespoon of Himalayan sea salt
- Pour in 2L of reverse osmosis filtered water and cover
- Leave for about 24 hours and remove the lemon pieces
- Drink about a pint of this water every morning on an empty stomach, before you have consumed anything else

Another method for this is to use half a teaspoon of sodium bicarbonate in a cup of hot water

9.7.3 MORE ABOUT WATER

- You might want to look into 'structured' water, and you can get vortex filters for this purpose. This is the closest to natural rock-permeated spring water as you can simulate, and I have seen direct examples of how this water can be used to grow enormous melons, or other fruit and vegetables, for example

A Theory of Everything

o When you observe water in a glass you may remember being taught about the meniscus that you can see on the top at the periphery of the glass around the edge, where the water appears to rise slightly above the level of the bulk of its surface. This is where at the surface of the vessel, the water is restructured as a result of the surface tension of the glass or ceramic. The key is in the surface area that the body of water is exposed to. So as in the case of the glass vessel, for example, only the water exposed to the sides of the vessel is restructured, and as a result of the volume to surface area ratio of the water in the vessel, only a minute portion of the water is restructured

o Now if you imagine water percolating through very narrow channels of porous rock as it is naturally filtered in nature, you can see that the surface area that the body of water is exposed to is greatly increased, and therefore restructuring most of, if not all of the water, and not just some of it. This is why natural spring water (not to be confused with products that are advertised as 'natural spring water' that are actually not) is so good for you!

- Another commonly advocated method is to drink hot lemon water, where you simply slice some lemon and put it into hot water to drink

- I also use, and do advocate the alkaline water drinking vessel provided by DYLN (tinyurl.com/5h8fdcwy) and an Eco-Bottle by Drink Safe (drinksafe-systems.co.uk)

- If you have no access to filtered water, then it is reported that there are good benefits from drinking about a pint of the cleanest water you have available to you immediately in the morning before you consume anything

9.7.4 HOW MUCH WATER SHOULD YOU DRINK?

It is recommended that you drink the equivalent of 9 to 13 cups, or 8 regular glasses of clean filtered water (or alkali water) every morning on an empty stomach (I drink a pint of alkali water every morning)

- The benefits of drinking water on an empty stomach in the morning are as follows:
- Rehydrates the body
- Increases your level of alertness
- Helps fuel your brain
- Fights sicknesses and strengthens the immune system
- Helps get rid of the toxins in your body
- Jump-starts your metabolism
- Reinforces healthy weight loss
- Improves complexion and skin radiance
- Prevents kidney stones and protects your colon and bladder from infections
- Promotes the growth of healthy hair

Water is very important to our body for many reasons; however, over-drinking can cause some slight or dangerous side effects such as:

- Feeling of constant fatigue
- Swelling of the feet, arms, legs, lips. This is a sign that you're experiencing an electrolyte imbalance
- Experience of a lingering headache
- Shaky or weak muscles, spasms, or cramps
- Disrupts your sleep during the night to urinate. The advice is to empty your bladder before sleeping

Reference: Drink Water First in the Morning (tinyurl.com/42zs26rn)

9.8 The Pineal Gland (Crown Chakra)

The pineal gland, which is apparently the gateway of our mind into the spirit realm, is rich in fluoride already. However, by subjecting it to too much fluoride you risk over-filling it and hardening its walls, and thus making it less effective, and reducing your spiritual, or energetic capabilities.

Ref: 50 Reasons to Oppose Fluoridation (tinyurl.com/2p86ya76)

See points 13 – 17 for details on tooth decay studies, and all the other 45 points make a strong case that this chemical damages the brain and lowers the IQ (point 22 and 23, and point 25 on the pineal gland).

This is why I use fluoride-free toothpaste, and I have a 6-stage reverse osmosis re-mineralising filter for drinking water. Sodium chloride is also very poisonous for the body. It's the chemical that makes your tap water taste like water from a swimming pool.

You can apparently cleanse the pineal gland, also referred to as the third eye, the seat of your soul, or the crown chakra, by drinking organic cloudy apple cider vinegar, which tastes like shit, but which I drink regularly.

Moreover:

- Iodine, a natural mineral found in seaweed, helps to de-calcify and cleanse the pineal gland
- Sun-gazing stimulates the pineal gland. The best time to do this is when the sun is not too hot, like first thing in the morning, just as the sun is rising.
  Sun-gazing And The Benefits From It (tinyurl.com/m895h3vs)

- Clean nutrition (see my previous nutrition insight)
- Consuming spirulina and chlorella, two algae which contain substances to help detoxify the body including the pineal gland.
  - 11 Powerful Supplements to Detoxify Your Pineal Gland, Boost Brain Power and Increase Vitality (tinyurl.com/unhh83xh)
  - Spirulina and Chlorella Aid Heavy Metal Detox (tinyurl.com/yc6h5vft)
- Get plenty of quality sleep and sleep in total darkness. One of the main functions of the pineal gland is to produce melatonin, a hormone which governs the sleep and wake cycles of the body, and contributes to activating the gland. Secretion of melatonin is stimulated by darkness and inhibited by light.
- Blue light from mobile phones and computer screens affects the pineal gland by preventing the production of melatonin. If you suffer from insomnia or are a bad sleeper, make sure to switch off phones, laptops etc at least 1.5h before your normal bedtime. It can make all the difference!
- Fluoride also affects the production of melatonin
  Reference: Pineal Gland (tinyurl.com/2nfy26k5)

It's postulated that the pineal gland produces DMT, which is a natural hallucinogenic similar to LSD. It naturally occurs in most, if not all, life forms such as plants, animals and humans. In humans, DMT appears to be produced when we dream as well as in larger amounts right before our death. Experiments with exogenous DMT have been done with striking outcomes; this molecule appears to enable us to tap into other realms such as the realm of the infinite. A fascinating account on the function of and experience while under the influence of DMT can be read in:

Book Reference: *DMT: The Spirit Molecule: A Doctor's Revolutionary Research into the Biology of Near-Death and Mystical Experiences* by Rick Strassman (tinyurl.com/26x7eetb)

You can begin to see that we are multi-dimensional beings having the human experience.

This is an excellent short video, where I have taken some of the information above from:

Reference: How to Decalcify Your Pineal Gland, The Science of The Pineal Gland and Third Eye Activation (tinyurl.com/2p86ubcm)

These are some of the most comprehensive and profound resources that I've come across for working with your pineal gland, and equally, for explaining much of what I've written about in this book, but in much greater detail, and in precise and scientifically researched and proven studies:

Book Reference: *Becoming Supernatural: How Common People Are Doing the Uncommon* by Dr Joe Dispenza (tinyurl.com/j82c8utf)
Book Reference: *The Pineal Gland: Tuning in to Higher Dimensions of Time and Space* by Dr Joe Dispenza (tinyurl.com/r7s646n7)

I cannot impress upon you more the significant value of Joe Dispenza's work, and his affiliation with, for example, Mind Movies, and the Heart Math Institute. His work is significant, and I am sure that if you do the research and follow the guidelines and meditations, then it will change your lives in an exceptionally positive manner, beyond what you could have ever imagined previously.

Many of the natural remedies that I use require pure water to work most effectively. And frankly, it tastes better. I'd consider myself to be a water connoisseur, if that's possible. I can always taste the difference between water from different sources and regions, even if, for example, it's provided to me in the form of a cup of tea or coffee. As I mentioned before, I have a six-stage re-mineralising reverse osmosis filter in my house, which is excellent. I also have a cyclone filter, which re-structures the water, in a similar way that water is structured when it passes through a natural porous rock structure, as a result of the maximised surface area that the water is exposed to. Studies have shown that when water is used to grow vegetables, for example, which has been restructured in this way, they grow to huge proportions! My good friend Roger Davidson first introduced me to restructured water, and many other overlooked healing techniques, including Rife technology and Tesla Diathermy, which are kept underground by Big Pharma and the financial institutions.

But by learning about these facts I knew which path I wanted to follow in terms of my general diet, health and wellbeing, and in terms of reducing my pain. Your mother inspired me and helped me to pursue a number of alternative practices and remedies over the years to try to combat my constant pain. One of the remedies she introduced me to was homeopathy, which she used regularly for you two too.

9.9 Homeopathy

Homeopathy is a very interesting one for me, as I see it as something beyond placebo, and explained very nicely in the book I mentioned earlier on, *The Field* by Lynn McTaggart. In a nutshell, Lynn explains that what she learned was that homeopathy is possible because water has a memory. What homeopaths do is to take a sample of whatever ailment they are trying to treat, like cold sores, sties, colds, pain and many other physical and mental ailments, and put it into some water, then dilute that water by half, so many times that theoretically there couldn't possibly be anything residually remaining. This diluted water is then used to coat little white sugary balls. Sounds bonkers, and Lynn explains it far better than I've just done. My intention is to sow the seeds for you to do your own research in the future, if you want.

Again, I don't actually care how it works, or whether it's purely placebo, because I have tangible evidence to suggest that it does work. And that's all I'm really concerned about! For example, I was once treated for a combination of stress and a nasty case of acne that I was experiencing on my back. Once I'd followed the course of pills, the acne on my back disappeared immediately, having been there for over a year, and my life became easier and more productive, as a result of feeling mentally better about myself. I accept that the mental well-being is relatively intangible, but the acne disappearing was a very tangible result!

A few pieces of advice that I've come across is that for the best effect, you need to consider several things during the time you are taking homeopathic remedies. Firstly, you're not supposed to have coffee or mint whilst taking these remedies. Also, when you take them, make sure that you do not touch the little white balls as you may disturb the coating on the surface, which is the active ingredient. The best practice for taking them is to put the remedy under your tongue and let it dissolve. There are more receptors under your tongue, and the remedy will be introduced to your blood stream quicker and more efficiently.

9.10 Cancer

Well here's another big topic! I've explained my understanding of the global pharmaceutical industry, and my take on common 'modern' diseases, and can say quite openly that the cancer 'industry' is a multi-million, if not, multi-billion-dollar industry. In addition to that I can say that according to the Cancer Act 1939, I am not allowed to say that there are other methods besides chemo or radiotherapy to cure cancer, so I will tell you that I am aware of many technologies and natural remedies which appear to purport to being able to treat such conditions effectively, and without the significant ramifications to health associated with radiotherapy. I point you towards this eye-opening 9-part documentary series:

Reference: The truth about cancer – episode 1 (tinyurl.com/2p93b94n)

9.11 Colloidal Silver

A good remedy that I've come across, which was introduced to me by my good friend Roger Davidson, is colloidal silver. There is also colloidal gold and colloidal copper, which provide different healing properties. Colloidal silver was used in hospitals around the world primarily as an antibiotic, until the pharmaceutical industry banned it because it's cheap to produce, cannot be patented, and because it works.

I've used it successfully and have seen many patients entertaining its benefits, including with assistance with dental issues, where the liquid is swilled around the mouth for a period to reduce inflammation and to address bacterial infection and root issues. I've also had exceptional results with dermatological issues, such as curing styes and cold sores, which would usually be with me for months, where I've been able to cure them overnight using my colloidal silver.

For example, according to Dr Axe (see link below), here are some uses for colloidal silver:

- Antibacterial and antimicrobial
- Wound care/skin health

- Pink eye/ear infections
- Antiviral
- Anti-inflammatory
- Sinusitis
- Cold/flu
- Pneumonia

His text is an excellent resource that explains the benefits of the medicine, and also explains the dangers of over-using the product, and that you must be careful of the source, and definitely worth a read, if you're interested:

Reference: Proven Colloidal Silver Benefits Or An Unsafe Hoax? (tinyurl.com/37hsxxuu)

Please also see:
Reference: What Is Colloidal Silver? (tinyurl.com/mx5bwjc9)

And this is where I have previously purchased colloidal silver, before I started to make it myself:
Reference: UKColloidalSilver

9.12 Meditation and Being in the Present

I could list a plethora of other natural remedies, alternative therapies and practices that I've come across and have used effectively. One of the most effective ones, which costs no money and requires no external person or therapy, is meditation.

I've mentioned the main benefits earlier, but I can't impress upon you the great benefits of this practice to your health, mental stability, and general wellbeing, particularly when starting at around your age (about 10 years old).

A Theory of Everything

I've developed a technique whereby I can eliminate my pain in an instant by using the principles that I've learned in this arena, and through my practice of Reiki. One method I use is a neat technique that my very good friend Alistair Savage taught me many years ago. The process is to take three deep breaths and relax, and consider the pain where it exists in your body to every extreme, and to accept it and allow it consciously, and to send it love. In this way I can remove any pain almost instantly, when I need to.

Here are some references to make a start with meditation:

Reference: How to Meditate for Beginners (tinyurl.com/bddv6kry)
Reference: How to Meditate for Beginners – WikiHow (tinyurl.com/yp8j7rbx)
Reference: 5 Meditation Tips for Beginners (tinyurl.com/mr24bxpd)

If you feel you don't have time for meditation, then regard this: Meditation can be as simple as the practice of being present in any moment. Not thinking about the past or the future, you are meditating. Being in the present can be practised anywhere, at any time and for any length of time. Do the dishes, for example, and be fully present with what you're doing, not allowing your thought to judge or think of past or future events. You'll find that this will remove a lot of the negative emotions we attach to past and future events, and eventually brings you more peace of mind.

More on this type of meditative practice can be found in great books such as:

Book Reference: *A New Earth*: Eckhart Tolle (tinyurl.com/5n7t88j3)
Book Reference: *The Miracle of Mindfulness*: Thich Nhat Hahn (tinyurl.com/ptushca3)

9.13 Exercise

That kind of goes without saying, doesn't it? And it's not necessarily such an issue for you guys right now, but the measured effects of exercise are profound, including actually giving us more energy, and keeping us healthier and more buoyant. However, in my state of mobility, I'd have to agree that I'm probably the last person you'd want to talk to about exercise!

9.14 Grounding

Walk bare foot on the earth as often as you can.

The body's voltage state:
- o The body's natural potential is at -0.25V and is affected by electromagnetic radiation
- o You can lower your voltage by grounding yourself by walking bare foot on the earth, for example (please see below for details)
- o This is thought to be of major significance in the health of the human body, as this action replenishes electrons and enables white blood cells to heal properly. It thereby helps to reduce inflammation, and the risk of developing inflammatory diseases such as cancer, etc.

Reference: The effects of grounding (earthing) on inflammation, the immune response, wound healing, and prevention and treatment of chronic inflammatory and autoimmune diseases (tinyurl.com/2p99r9a8)
Reference: The Science of Grounding (tinyurl.com/2p9c3z4s)
Reference: Earthing.Com (www.earthing.com/)

9.15 Being in Nature

I cannot embellish enough just how important and beneficial it is for us as human beings to spend regular time in nature! The benefits are profound on so many levels, and equally in stilling and quietening the mind, and in connecting with yourself and the universe. So, yes, go naked on the beach (where permittable), wander through ancient woodland, hike up sacred mountains, marvel at the awesomeness of nature, and hug some trees!!!

I used to scoff at the *happy clapper tree hugger wankers*, but now I have a deep respect for people who understand the significance of such activities. You see, I've learned that the trees, and indeed all plant life, are conscious beings, with organised thought, albeit in a very different sense to what we're used to, and that by hugging trees, you are not only absorbing electrons from the earth in a grounding sense, but you can also connect to the vast knowledge and wisdom that they hold!

9.16 Sleep

Getting enough sleep is paramount to your existence on earth. It is not an optional lifestyle luxury, it is a non-negotiable biologically necessary condition of living, and when we do not get enough sleep, as society in industrialised nations pushes us further and further towards, the silent effects can be seen emanated in an inability to store new memories, or memory loss, it can distort your genes and DNA genetic code, for example, promoting tumours or raising your stress levels, and is even associated with so-called diseases like cardiovascular disease, and Alzheimer's.

A Theory of Everything

Please see the following TED talks by Matthew Walker for information:

Reference: The Sleep Deprivation Epidemic with Matthew Walker (tinyurl.com/yckm4rdt)
Reference: Sleep is your superpower | Matt Walker (tinyurl.com/2p9yh5cf)

Please see the following basic advice for protecting yourself:

1. Go to bed at the same time and wake up at the same time every day, regardless of whether it's a weekend or not
2. Sleep in a cool, well-ventilated environment. You will always fall asleep and sleep better if it is too cold, compared to if it is too warm. Ideally your sleeping environment will be at around 18 degrees C, or 68 degrees F
3. Switch off all electronic devices at least 1 hour prior to going to bed. No matter how important your work or entertainment, your sleep is FAR more important, trust me
4. Avoid drinking liquids after 7pm in the evening as this can disrupt your sleep by the need to urinate during the night. Make sure to get your daily water intake in, prior to that time
5. Do a sleep meditation or practice some mindfulness for 10-20 minutes once you have settled in bed. This can make all the difference between an anxiety-ridden night or sound sleep
6. Avoid caffeine, alcohol and sugar completely if you are a sensitive sleeper
7. Avoid bright lights. Insomniacs may find it helpful to use candlelight only
8. Avoid vigorous exercise before bed. Exercise should be done latest 4h prior to sleep
9. Remove all electronic devices from your bedroom and switch off Wi-Fi equipment

10. Install a grounding sheet in your bed
11. Trust your body. Where possible and practical, if your body tells you it's tired, then allow it to sleep

9.17 Reiki

I've mentioned this before several times, and here is a brief introduction to Reiki from my Reiki Master and Goddess, Jill Sudbury:

Reiki means universal Life Force energy. When we're connected to this source of all love, power & healing light, we can transform ourselves & those around us, being ever mindful of our true nature & search for meaning. When you learn this wonderful energy tool, you become connected to the source of life, thus helping you to remember your true self & to flow with the divine energy of love, life & everything.

This is a creative process, the side effect being that you can share it with others through your presence, higher vibration, or by giving a Reiki session. It can be used for plants, animals, nature, everything! And to cap it all...it is not a religion. It can be combined with any belief as long as that is for the highest good of the self or those around you.

I've had the pleasure of teaching Reiki for over 22 years and have enjoyed very much seeing the empowerment, meaning and lights go on in my students. Yes, anyone can learn.

It was taught in this form by Dr Mikaou Usui in Japan in the 1920s. It's survived wars & cultural differences, becoming a valuable life tool in many walks of life around the world. The principals being complete mindfulness.

Just for today I shall not worry, not get angry, will work honestly, will respect all life & I shall count all My Blessings.

Dr Usui's memorial stone states 'Reiki will conquer the world, and heal not only its inhabitants but also the Earth itself '
Jill Sudbury

Reference: Jill Sudbury.com (tinyurl.com/y323t6kb)

9.18: Life's Best Free Healing Sources

Here's a list of the best-known free sources of energy, healing and vitality available to everyone:

- Sunshine – sun gazing
- Air – deep breathing
- Exercise – stretching
- Water – pure and restructured, if possible
- Diet – mostly organic, plant-based or foraged
- Rest – proper sleep, and when your body requests it
- Play – play with an innocence that forgets the worries of the world
- Laughter – long, hard and often
- Meditation – be present in the now as often as you can

"I love people who make me laugh. I honestly think it's the thing I like most, to laugh. It cures a multitude of ills. It's probably the most important thing in a person." Audrey Hepburn

---

Thanks Very Much and Best of Tuck with It!

And that's pretty much where I'm at with my life advice right now. I'm still learning and developing my knowledge, experiences, techniques and philosophies, and I want to learn from you, or anybody else who comes into my life for this purpose. I want to keep an open mind, and more importantly, I want to help you, and anyone else who's interested, with what I believe I've learned about life so far, if it's what they are open to, and in the process of finding.

A Theory of Everything

"We cannot solve problems by using the same kind of thinking we used when we created them" Albert Einstein

"None of us are getting out of here alive, so please stop treating yourself like an afterthought. Eat the delicious food. Walk in the sunshine. Jump in the ocean. Say the truth that you're carrying in your heart like hidden treasure. Be silly. Be kind. Be weird. There is no time for anything else" Anthony Hopkins

"Accept – then act. Whatever the present moment contains, accept it as if you had chosen it. Always work with it, not against it… This will transform your whole life." Eckhart Tolle

Appendix 1
Human Psychology and Biology

I wanted to capture something that inspired me in the movie *Zeitgeist III*. It backs up and expands upon everything I've learned about human psychology. Especially with regard to how our environment shapes us from even before we are born, and blueprints our beliefs and behaviours. It also explains some of what was missing in my understanding, and that is, how much of our beliefs and behaviour is inherited or 'genetic'. Pretty much nothing of it, in fact. Anyway, here are my notes from the documentary:

Nature versus Nurture. We have a totally false understanding of what makes us who we are, at a fundamental level. It is virtually impossible to understand how biology works, outside the context of environment. Behaviour is taught, or understood by most, to be genetic. And therefore, negative behavioural traits are not addressed positively in society, because it is a commonly understood belief that this is not possible. This is nonsense (Dr Robert Sapolsky – Professor of Neurological Sciences, Stanford University). It is widely thought that conditions like ADHD or schizophrenia are genetically programmed. The truth is the opposite. Nothing is genetically programmed. There are only a few very rare diseases that are truly genetically programmed. Most complex conditions might have a predisposition with a genetic component, but a predisposition is not the same as a predetermination. So, looking for predetermined diseases in genomes is an utter waste of time, because most diseases are not controlled by our genes. Cancer, strokes, rheumatoid conditions, autoimmune conditions in general, mental health conditions and addictions, none of these are genetically inherited. Breast cancer, for example, out of 100 women, only seven will carry the breast cancer genes; 93 do not. And out of 100 women who do have the genes, certainly not all of them get cancer (Dr Gabor Maté – physician, author, Portland Society).

Genes are not just a thing that make us behave in a particular way regardless of our environment. Genes give us different ways of responding to our environment, and it looks as if some of the early childhood influences affect gene expression, actually turning on and off different genes. This puts you on a different developmental track, which may suit the kind of world you've got to deal with (Professor Richard Wilkinson – Professor Emeritus of Social Epidemiology, University of Nottingham).

It has been understood categorically that in representative studies, for people who suffered abuse as children, the abuse actually caused a genetic change in the brain. This is described as an epigenetic effect (epi means on top of), where whatever happens environmentally will either activate or deactivate certain genes.

A study was done in New Zealand, where a few thousand individuals were studied from birth, right up until their mid-twenties. They found that they could identify a genetic mutation, an abnormal gene, which did have some relation to the predisposition to commit violence. But ONLY if the individual had also been subject to severe child abuse. A child with this abnormal gene would be no more likely to be violent than anybody else, and in fact actually had a lower rate of violence, as long as they weren't abused as children.

You cannot genetically modify anything successfully in the absence of environmental impact. If you remove the 'learning' gene and then subject the animal (because the test was done on mice) to a healthy and positive, enriched and stimulating environment, then very quickly the animal will rapidly overcome that deficit. So, the actuality is that there is a genetic contribution to how this organism responds to environment. Genes may influence the readiness with which an organism will deal with a certain environmental challenge. This is very far from the general public understanding that 'it's genetic', and is completely debilitating to social understanding of pretty much anything that doesn't fit into the idea of how society should be (Dr James Gilligan – Former Director at the Center for the Study of Violence, Harvard Medical School).

If you believe that 'it's genetic' then it's easy to see how you might succumb to the fact that there is nothing that you can do to change the predisposition people have to become violent. If you believe that it's genetic, then it's easy to see how you might believe that all you can do if somebody becomes violent is to punish them by locking them up, and not worry about changing the social environment and social preconditions that may have led to this person becoming violent, because you will see that as being irrelevant.

The genetic argument is an excuse for allowing us to ignore the social, economic and political factors that underlie many troublesome behaviours.

Addictions: No substance and no drug is by itself addictive, and no behaviour is by itself addictive. The real issue is what makes people susceptible. To understand what makes people susceptible, you have to look at their life experience. Addictions are NOT genetically inherited!

It has been shown that if you stress mothers during pregnancy, then their children are more likely to have traits that predispose them to addictions, because development is shaped by their psychological and social environment. The biology of human beings is very much affected by and programmed by their life experiences being in the uterus. Environment does not begin at birth. Environment begins as soon as you are a foetus. Prenatal effects have a huge impact upon the future behaviour of a developing human being.

Memory: There are two types: *Explicit* memory – this is recall, when you can call back memories. However, the hippocampus structure in the brain that enables us to store explicit memory does not develop until about the age of eighteen months. Before that there is *Implicit* memory, which is an emotional memory, where the emotional impact and the interpretation that the child makes of those emotional experiences are ingrained in the brain in the form of nerve circuits ready to fire, without specific recall.

D. W. Winnicott (an English paediatrician) said that fundamentally two things can go wrong in childhood: one is that something that isn't meant to happen, happens; and the other is that something that is meant to happen, doesn't.

Traumatic, abusive or abandonment experiences are examples of the first. The second is experienced as a result of lack of love and affection provided by the parents as the child grows up. Not abuse or neglect or trauma, but the presence of the emotionally available nurturing parent is not available to them because of the stresses imposed upon the parent in our society. Allan Schore calls this Proximal Abandonment, when a parent is physically present, but emotionally absent.

A study showed that the most violent criminals in the US had themselves been victims of the most appalling and unimaginable child abuse, and that the murderers were often the survivors of their own attempted murder, at the hands of their parents, or other people in their social environment, or were the survivors where other family members had been killed.

Buddha said that everything depends upon everything else; that the *one* contains the many, and the *many* contain the one. You cannot explain anything in isolation from its environment.

The *Bio Psycho Social* nature of human development states that the biology of human beings depends very much upon their interaction with their social and psychological environment. Daniel Siegel of the University of California developed the phrase, *Interpersonal Neurobiology*, which means that the way that our nervous system functions depends very much on our personal relationships, in the first place with the parent and caregivers, and in the second place with other important attachment figures in our lives, and in the third place with our entire culture.

Humans have spent the majority of their existence in this current phase of human evolution as hunter-gatherer groups, with loving and caring communities of family and friends, and in societies with few possessions, no enforced laws, and very little violence. Organised group violence was something that just didn't occur during these times. Today, there is a huge variation in the amount of violence observed in different societies globally. There are some societies that experience virtually no violence. There are others that destroy themselves. Some of the anti-Baptist religious groups, such as the Amish, the Mennonites and the Hutterites have no recorded cases of homicide in their history. During the wars, these people would rather go to prison than serve in the Army. In the kibbutz community in Israel, authorities will actually send offenders there to be cured, because it teaches them how to live a non-violent life.

Human Nature Myth: People are competitive by nature; they are individualistic; they are selfish. The reality is quite the opposite. The only way to talk about human nature concretely is by recognising that we have certain human needs: companionship and close contact. To be loved, to be attached to, to be accepted and to be received for who we are. If these needs are met, then we develop into people who are compassionate and cooperative, and with empathy for other people. The opposite is what we often see in society, which is the distortion of human nature, because so few people have those needs met. This is no clearer than in the observation that, if a newborn baby is not touched, it will die.

Violence is more prevalent in unequal societies.

The health implication on poverty is not about being poor. It's about *feeling* poor.

Everything is statistically better in more equal countries.

Appendix 2
FELT – Dr Linda Mallory
(Child Psychologist and Author)

Our thoughts play a key role in relationships with our emotional wellbeing.

I have created FELT as an approach to help teachers, parents and children understand the relationship between our thoughts and feelings. This in turn

reduces stress and anxiety and optimises a calm school and home environment.

We are living in the feeling of our thoughts. FELT focusses on the practice of being aware of our thoughts by encouraging us to notice our feelings (F), what we enjoy (E), what we learn (L), and what we are thankful for (T).

FELT is an integrated approach which uses positive and humanistic psychology to help to reduce stress and anxiety. In my work with children, school staff and parents, I have used FELT and seen the benefits of this simple daily practice to increase our clarity of thinking and reduce stress.

Contrary to popular belief wellbeing is not about just being happy. FELT is about noticing and acknowledging our full range of thoughts and feelings, and developing an awareness of our innate emotional wellbeing.

A Theory of Everything

As humans we feel our thinking and there are three principles we all share. We have thoughts, we have a mind, and we have consciousness or awareness.

Supporting children with this understanding helps awareness of their own emotional wellbeing. Knowing that we feel our thinking and have thoughts that come and go can help return us to a natural wellbeing and calm state.

Stress and anxiety do not come from what is going on in our lives; they come from our thoughts about the situation.

When we judge our thinking, this leads to stress and anxiety. Acknowledging our thoughts without labelling them as 'good' or 'bad', 'right' or 'wrong', 'better' or 'worse', helps us to have clarity and we are more likely to have another thought and be creative, productive and reach our potential.

Noticing the energy or emotion behind a thought (emotion being energy in motion) helps us to be more aware and feel our emotions. They are telling us something about our thinking.

The kinesiologist David Hawkins in his research (power versus force) concluded that emotions have an energy.

Low-energy emotions are sadness (red), fear (orange) and anger (yellow).

Higher-energy emotions being love (green), joy (blue) and peace (purple).

The lowest form of energy emotion being shame and guilt, and the highest form of energy emotion being gratitude and compassion, corresponding with the colour spectrum.

The FELT feelings wheel has been created to produce a tool to help notice our emotions without judging them as being good or bad, positive or negative.

A Theory of Everything

FELT feelings wheel:

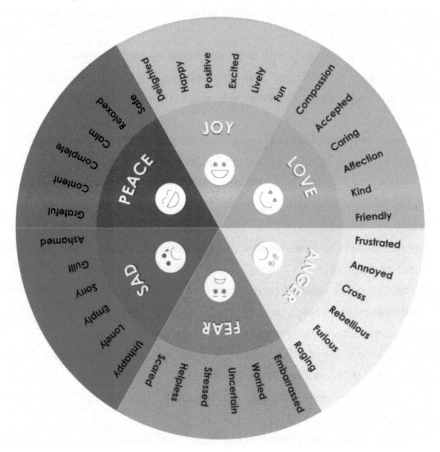

Having feelings makes us human. Having a range of emotions is normal and an awareness of feelings helps to develop emotional literacy. One type of emotion is not better or worse than any other, and the range of different feelings is a healthy expression of being human.

When we judge our emotions this can lead to feelings of shame and guilt which is a low energy compared to high-energy emotions such as peace and gratitude.

A Theory of Everything

The most important aspect of the feelings wheel and FELT is that it is a tool to accept and allow our full range of emotions. We are humans, we have thoughts, and we are aware of them. They come and they go. If we don't allow thoughts and emotions to flow and we repress or suppress them the energy can be blocked and children learn to hide their emotions which can over time lead to mental health needs.

Appendix 3
What They Don't Teach You
at School

- **What life's really all about, or at least the desire to think about it, and embrace the facts in order to define your own truth**

It is clear that not everybody is ready for this level of deep critical thinking, and I accept that you might not be either, but please be conscious of the fact that it's understood that society deliberately dissuades us from this level of thinking, and to me, quite obviously because if people worked out what it was all about, or *'awakened'*, then that would threaten the entire global financial institution, and the control over the mechanisms behind the reasonably organised, and progressively tyrannical society in which we live.

- **The importance of the language that you use in communication or conversation (a study of which, for example, is referred to as neurolinguistic programming, or NLP)**

This is of significant importance in your life, not just for the effectiveness of your communication with others, but also for the way you think, and the way you view life and what subsequently happens to you. I have described my understanding of this previously in section 7.5.

• How to think

It is clearly understood that for as much as your teachers try to teach you this skill, the global schooling system is derived and orchestrated at the highest level to try to erode your ability to think for yourself, and to think critically. This is a well-documented topic.

• Emotional awareness and intelligence, or wisdom

"We learn plenty about our physical health, but what about our emotional and mental health? What about our inner worlds? Could there be any topic more relevant to students and young adults than understanding and managing their stress, anxiety, and emotions? If mindfulness and emotional awareness was as essential to the public school curriculum as Common Core math strategies, we just might raise a healthier generation of humans"
Babble.com

• How to be a parent

I'm very grateful to your mother for the amount of research she did in the early days of our parenthood, and for the books that she studied and recommended we read together, in an attempt to give you the best start we could. For example, when you were born, Alex, she made sure that we mitigated any potential animosity or envy from you, William, by talking to you about how great it was going to be for you to have a little brother. And then, when Alex was born, there was a present to you from him. I strongly believe that this had a hugely positive impact upon the strong bond that you have in general, even though I'm sure you will beg to disagree at this stage in your life, with all the challenges that come with being brothers!

I've equally come to realise that being a parent is as much about being able to address your child's psychology as it is in being able to observe, accept, and address your own psychology, and your own core beliefs and subsequent behaviour.

One of the many exceptionally insightful and useful books that I've had the pleasure of reading on the topic of parenting, and a whole lot of other life advice, is as follows:

Book Reference: THE GROWN-UP'S GUIDE TO TEENAGE HUMANS: How to Decode Their Behavior, Develop Trust, and Raise a Respectable Adult by Josh Shipp (tinyurl.com/yc2p2jdn)

- **That much of our taught and socially accepted understanding of the world and how the Universe works is based on 'theory' only**

NEAT LIFE HACKS: REMEMBER THAT WHEN ANYTHING IS PRESENTED AS A THEORY IT IS REFERRED TO IT AS SUCH BECAUSE THAT'S *EXACTLY* WHAT IT IS, A *THEORY*! FOR EXAMPLE, THE THEORY OF RELATIVITY IS A THEORY. DARWIN'S EXPLANATION OF EVOLUTION IS A THEORY. GRAVITY IS A THEORY. PARTICLE PHYSICS AND PARTICLE GEOMETRY ARE A THEORY. "THERE ARE NO PARTICLES, JUST RESONANCE AND FREQUENCY" – SEE DOUBLE-SLIT EXPERIMENT! QUANTUM THEORY AND HELIOCENTRISM, THE THEORY THAT THE EARTH AND PLANETS REVOLVE AROUND THE SUN – NOW THAT'S A GOOD ONE! ETC...

- **The importance of travel, and putting yourself outside of your comfort zone**

"WHETHER YOU LEAVE THE COUNTRY OR JUST LEAVE YOUR TOWN, TRY TO PLAN AT LEAST ONE TRIP EVERY SIX MONTHS. USE WEBSITES LIKE STUDENT UNIVERSE (TINYURL.COM/2PNDXZ) TO FIND AFFORDABLE DEALS, AND COMMUNICATE WITH YOUR UNIVERSITY ABOUT TRAVEL GRANTS. CONSIDER GOING ON A GAP YEAR. ACCORDING TO AN INTERVIEW (TINYURL.COM/YBU59BZC) WITH DALE STEPHENS, THE FOUNDER OF UNCOLLEGE GAP YEAR (UNCOLLEGE.COM/PROGRAM), STUDENTS THAT TAKE A GAP YEAR HAVE HIGHER GRADUATION RATES AND GPAs THAN STUDENTS WHO OPT NOT TO TAKE ONE" UNCOLLEGE.ORG

One of the highest formal education bodies in the US, Harvard University, who churn out some of the best and most influential graduates, make it a prerequisite that all students they take on have had a gap year before they commence their further education.

- **How to handle money or your finances**

"BE THAT IMPRESSIVE 20-SOMETHING WHO KNOWS THE INS AND OUTS OF THEIR BANK ACCOUNT, THEIR CREDIT SCORE, AND THEIR INVESTMENTS. CREATE AN ACCOUNT AT MINT.COM (MINT.COM) AND START A REALISTIC BUDGET FOR YOURSELF. SET ASIDE 10% OF EVERY PAY CHECK YOU GET, AND LOOK OUT FOR FREE MOOCs ABOUT PERSONAL FINANCE. (TINYURL.COM/Y7OJSK4Q) GET FAMILIAR WITH HELPFUL MONEY-SAVING BLOGS LIKE 20SOMETHINGFINANCE (20SOMETHINGFINANCE.COM/), THE BILLFOLD (THEBILLFOLD.COM) AND THE FINANCIAL DIET (THEFINANCIALDIET.COM)" UNCOLLEGE.ORG

Book Reference: The Richest Man in Babylon – George Clason (tinyurl.com/y9w42hyv)

• About love or relationships

"Remember that time spent on your relationships is not time wasted, even if those relationships eventually end. Every friendship and relationship you form can teach you how to strike the right balance in a life partnership. Write down the qualities you are looking for in a partner, and focus on the qualities you have to offer. Make sure to protect time each week to spend on your relationships: don't let yourself become a one-sided person"
UnCollege.Org

"We take classes and a test before getting a driver's license. We take Lord knows how many exams before getting into college. We're even offered a variety of parenting/birthing/ breastfeeding classes before having a baby. And yet I could walk into a courthouse with a simple registration and some makeshift rings and call it a marriage. How can something so complicated and important – something that affects everything from our money to our health to our happiness – have next to no training or instructions? This is another thing that should be learned at home in theory, except many kids have really crappy relationship role models because their parents had crappy role models because THERE'S NO EDUCATION ON MAINTAINING RELATIONSHIPS" Babble.com

---

• How to run a business, or how to build a career that's all your own

This is something I can help you with, and believe that I have much experience in, if you want it – if nothing else, I've certainly made enough mistakes to have had a wealth of positive learning experiences!

"Take time to determine what puts you in 'flow.'(tinyurl.com/myu7k28) Take classes outside of your subject, and think creatively about ways that you can make a living. And be sure to check out the UnCollege blog (tinyurl.com/y95eowvd) for more tips about pursuing your passions and building a singular, extraordinary life" UnCollege.Org

---

- **About the government**

What can I say? Can of worms... And the topic for another book entirely, embellishing much of what I have alluded to in this book!

- **About History**

Jeese Louise, this is also a massive can of worms! If you do your research, you'll find a plethora of so-called 'conspiracy theories' all suggesting that our true human history is heavily doctored and biased, covered up, and at worst, completely lied about and shrouded from our view.

"Yes, I'm a conspiracy nut, because it's better than being a brainwashed sheep" Jack Nicholson

---

For example, some of the topics that are well documented if you look for them, ask questions like:

- Who built the pyramids and what were they for? It's made very clear, from the perspective of our current understanding of physics and how the world works, that the ancient Egyptians could not have built the pyramids, or sphinx, or other ancient structures using the documented technology that they possessed at the time. It's explained very scientifically and very convincingly by experts at the top of their fields, that it just wouldn't have been possible, and that the given explanations that are taught, could not possibly be true.

- On the same theme, who built the megalithic structures around the globe? Again, it appears to be well understood scientifically that nothing in our taught history can explain how these structures were created, or what for. It is scientifically proven that even with the technology that we possess today, we couldn't possibly construct these structures and buildings that we see all round the world. Someone else must have built them, and it is thought that they were built far longer ago than we are taught or led to believe. For example:

  - The Unfinished Obelisk of Aswan, Egypt
  - Carnac Stones, France
  - The Dolmens (or passage mounds) of Antequera, Spain
  - Ggantija, Malta
  - Stone Spheres, Costa Rica
  - The Olmec Heads, Mexico
  - Yonaguni Monument, Japan
  - Gulf of Cambay, India
  - Moai, Easter Island
  - Göbekli Tepe, Turkey
  - The Bosnian Pyramids (referenced below)
  - Avebury and Stonehenge, UK

Documentary Reference: Mysterious Prehistoric Sites from Around the World (tinyurl.com/y6vrk4v9)

A Theory of Everything

Documentary Reference: Stunning Bosnian Pyramid Facts That Prove They Are Real Pyramids (youtube.com/watch?v=vDM8bYxygy0)
Documentary Reference: Nikola Tesla – Limitless Energy & the Pyramids of Egypt (youtu.be/Ft1waA3p2_w)

Please see an excellently presented and concise overview of the situation:

Documentary Reference: Evidence at Giza just keeps stacking up! More NEW Evidence!! (youtu.be/WRjQxoXi7tY)

- Other documented questions include:

  - What's the truth behind the intricately constructed and arguably non-man-made crop circles that appear, and their actual origins?
  - What is actually situated in Antarctica, and why has there been apparently so much interest in the region since around WWII?
  - What flying technology do global authorities actually have access to? What do the global elite actually know about aliens? Are 'The Greys' really aliens? Is there a space station on Mars, and was the documentation of the 'first' moon landings actually the first time man had landed on the moon?
  - What is Planet X, or Nibiru? What of the history of the Anunnaki and Sumerians?
  - Why is it reported that seemingly groundbreaking, history-changing findings are covered up and swept under the carpet, such as reported findings of artefacts from civilisations that far predate documented, taught, or accepted history?
  - Who really funded the world wars, and what were they actually for?
  - Who does the CIA work for?
  - Why do they really put fluoride in our water?

- What of the Klerksdorp Spheres that are believed to be 'man'-made, and billions of years old?
- What of SASER technology, that can project sound energy at a speed faster than the speed of light?
- What of Torus and Cone Stones?
- What of the significance of sound and magnetism?
- What of the occult?
- What of gravity, which is observed, but not really understood?
- What of one of the most extensive genocides ever observed in recent times, that of the extermination of the Native Americans, and also that of the Australian Aboriginals. And equally, any other atrocity perpetrated by even our own country in recent history, that we don't get to learn about in school. Why? Well, either it's all just conspiracy theory, or, because if we knew and accepted 'the truth', then who knows what would happen to society as we know it?
- What's the actual agenda and truth behind 5G and electromagnetic radiation (EMR)? Or the smart meter, for that matter?
- What of the significance and potential power of the pineal gland?

These are just questions, and only some of them. In our society, we're taught to view such questions as being foolish or crazy, and taught that conspiracy theorists are crazy fools. And maybe some of them are. And maybe all of them are. But the question I'm most interested in is, whether we allow ourselves to think about such things as a result, and whether we do the research for ourselves, and what we trust to be our truth as a result of that research.

I don't know the answers, and I wouldn't suggest that I know for sure what the truth is on any of these topics, and I'm acutely aware that it's a niche market. Conspiracy sells, and it is big business.

However, I am convinced of one thing, and that's the propensity for governments (or governing bodies) to lie about, and to cover up anything that threatens to get in the way of their primary objective, and that is to make money, or enhance the financial institution that governs them, and to control the general population.

This is a well-documented understanding that is published openly in the public domain. For example, it's common to find that information is released into the public domain after many years (usually around 50) surrounding atrocities that turn out to have been perpetrated by leaders against their own people. It would appear that this is mitigated by the current leaders in several ways. Firstly, because our so-called 'democracy' is orchestrated in a way that there's a regular turnover of rulers, such as the 4-5 year terms applied to US and UK government, so it is easy for them to blame and shame the specific people concerned with the exposed wrongdoing, and to brush it under the carpet. And secondly, I believe that full advantage is taken of the fact that we human beings are in general incredibly gullible and our minds are influenced so easily. We inherently tend to have very poor memories, and do not have the capacity to amass all of the reports that we've ever been exposed to, and see clearly that history is repeating itself, time and time again, even if at a higher level, we're aware of this fact in general.

Many of the documentaries that I've watched have put things into perspective by chronologically listing the events and the exposure of such events, and by explaining how public opinion has been swayed towards conspiracy theory, and away from the severely damaging truth.

Equally, I can understand that anyone in any profession, who has built up a presence and following in the scientific community, or any other specialism, wouldn't want their life's work and efforts being threatened by a theory, or finding, which might change the very foundations of their integrity.

- **How to focus**

"Scientists are now realizing that the newest crop of humans have an unprecedented ability to multitask, probably due to neuroplasticity (our brains' ability to adapt and change to the environment). New York magazine reported that kids can '[conduct] 34 conversations simultaneously across six different media, or pay attention to switching between attentional targets in a way that's been considered impossible.' But with the give comes the take, and studies show that these kids have less of an attention span than ever before. Perhaps the best thing we can teach these kids is to single-task, and to really listen and focus, rather than succumb to every distraction like a dog in a field of squirrels" Babble.com

---

- **How to survive without technology**

Just ask yourself, if, for whatever reason, electricity would be cut off for a day (how) would you get through that day?

- **How to survive in nature, grow or forage food, make a fire and how to purify water**

These are basic life skills and would come in handy if there is ever a lengthier electricity cut than just a day!

- **The dangers of watching TV and believing everything you watch or see in the media**

"What if I told you that television is the monster in your home, and it's called a program for a reason. Your television is nothing more than an electronic mind-altering device that has been designed to psychologically change the way you view reality"
Morgan Freeman

- ## How to buy a home or car, and how to do repairs

Males and females alike should be taught the basics of fixing in- and outside repairs, or at least know how to change a tyre!

- ## About the pitfalls of credit and debt

Historically, credit and debt were orchestrated initially by the creation of building societies, which produced the single greatest debt that many people will have, and that is the mortgage. Since this time, financial institutions seem to have leveraged this credit and debt situation in the form of credit cards, no-down-payment loans and, most recently, student loans, in order to further fuel the financial machine, which is rumoured to be currently spiralling out of control. I'm actively observing the whole 'Bitcoin' phenomenon with an open mind and eager anticipation!

- ## About the true dangers of alcohol and addiction

As a belligerent 'addict' for most of my life, and having finally and terminally beaten alcoholism, I find it perplexing that so-called 'addicts' are treated with such contempt in society. For example, it is explained by AA that alcoholism is a condition that you're born with, and it isn't until you take your first drop of alcohol that you begin the slippery slope to insanity and self-destruction, *or not*. Perplexing! I have a friend who works in the community (Emma Mullis), who argues passionately against this hypothesis, but essentially, we're both fighting for, and arguing about the same condition.

I've since learned that, actually, addictions are primarily conditions triggered and governed by early childhood traumas (see section 2.4 and Appendix 1).

What I'd also mention at this stage is that there are schools of thought which say that alcohol is legal in society because it nulls the minds of the individual, and it reduces the propensity for critical thinking. There is a reason that spirits are called a *spirit*.

- **About healthcare and the benefits of natural healing remedies and spiritual healing practices versus the well-documented, often poisonous and destructive effect of man-made medicines**

This is a topic that I've thoroughly researched for most of my adult life, beginning with my quest to relieve my constant pain, and the ineffectiveness of commercial mainstream medicine.

- **About the Bible, or Koran, or any of the sacred scripts**

My God (excuse the pun), there are some very worthwhile, beneficial, profound and fundamental teachings to be learned about life, and who we really are, in these texts. Unfortunately, the spiritual and underlying love that was once represented in these texts has been all but bastardised by a corruption-driven need to control the general population. It's rumoured that many texts have been deleted or altered accordingly in history. I mean, just look at what Hitler was able to achieve by blatantly and openly corrupting the texts and teachings of the Catholic Church in the build-up to, and during his reign in WWII, as an attempt to get public backing for his genocide and vision.

Documentary Reference: Zeitgeist I (II and III) (youtu.be/pTblu8Zeqp0)

- **Self-defence and first aid**

- **Learning from 'failure', and the importance of doing so**

"I've missed more than 9000 shots in my career. I've lost almost 300 games. 26 times, I've been trusted to take the game winning shot and missed. I've failed over and over and over again in my life. And that is why I succeed" Michael Jordan

"You have to fail early, you have to fail often, and you have to fail forward" Will Smith

"That's how you become a great man. Hang your balls out there" Jerry Maguire

---

- **Finding a job or time-management**

The corporate world of employment is changing rapidly, particularly with the advancement of robotics and automation, and there will be dramatic changes required in the very near future in the skills you are going to need, if you want to go down the path of playing the employment game.

- **The law, and how to operate in society effectively, but without compromising your life, or making things difficult for yourself;**

"A foolish faith in authority is the worst enemy of the truth"
Albert Einstein

- **How to avoid mental breakdowns or burning out**

"Start practising meditation for just ten minutes a day. Take advantage of on-campus counselling options. Set boundaries for when you allow yourself to check your inbox, and try to limit your 'screen time' in the mornings and evenings. Know how to ask for a 'personal day'. Professors and employers will respect you for knowing your boundaries" UnCollege.Org

---

- **How to bounce back, or the art of falling with grace**

"Read the biographies of people you admire, and make a note of the ways in which they coped with set-backs. Always remind yourself of the big picture, and learn not to sweat the small stuff (dontsweat.com)" UnCollege.Org

---

- **How to go through your teenage years, and what to expect:**

Book Reference: Passages: Predictable Crises of Adult Life – Gail Sheehy (tinyurl.com/yahhrsgd)

- **How to cope with problems related to mental health**

  Mental health problems are just as valid and worthy of attention as physical health problems, and in my opinion, grossly and blatantly misunderstood, misdiagnosed and covered up by mainstream 'medical' practice. I'm in fact utterly disgusted by the apparent misunderstanding of so-called mental health issues in society, and the propensity to, more often than not, prescribe expensive, chemically manufactured drugs to treat the issues, rather than focussing on preventative and psychological remedies and treatments.

- **How to network effectively (it's not so much about what you know, but who you know)**

"Try one-on-one networking with older students, professors, and others in your field. Don't be afraid to reach out with an unsolicited email. Remember, they were once in your position! And remember that serving as a connector – being able to link two friends together – is just as important as forging connections for yourself" UnCollege.Org

---

- **How to use your instinct or trust your gut**

  This is very much in line with my principle of making decisions quickly, and changing them slowly (Discussed in sections 4.3 and Parts 6 and 9)

"Start by making yourself a person who doesn't second-guess the choices you make, even small ones. Have to choose which restaurant to check out for dinner? Go with your first instinct. Don't know which class to sign up for? Go with the one that 'feels right.' And be sure to pay attention to when something feels 'off': trust your gut, and act immediately" UnCollege.Org

- **Any of the core life principles for harmony and happiness**

  We can observe such principles and a spiritual understanding of life as practised by indigenous peoples, tribes and many early societies, although it is gradually being corrupted even in Eastern societies.

- The amazing capability and significant power of the human mind

- About alchemy

- About the occult, about Freemasons and secret societies

- The truth about so-called 'drugs', narcotics and psychedelics

It's very clear to me, and a great many other people, that if 'drugs' is a term to represent something that is bad, or detrimental to health, or even evil, then this term would be most accurately represented when referring to mainstream pharmaceuticals, and not natural herbs and substances.

So I'll just come out and say it, and I'll provide a few references, too: if you do your own research into the vast benefits of natural (non-synthesised) herbal and plant-based 'drugs', then you'll find a plethora of information to support that if you want to find out who you truly are, then administration of such natural substances will not only expand your mind and consciousness and enable your awakening, but also assist in combating the toxins and poisons that we are routinely subject to through food and water, and electromagnetic radiation, for example. These substances even have the potential to cure so-called 'uncurable' diseases, such as cancer and other possibly deadly conditions associated with the financial giant that is the pharmaceutical industry.

Reference: Graham Hancock – The War on Consciousness – TED Talk (youtu.be/Y0c5nIvJH7w)
Reference: Cannabis: A Lost History (youtu.be/X2p6qFT_Zjg)
Reference: Psychedelica: Psychedelics And Consciousness (youtu.be/c-ErA_acTq8)

As a footnote, I will say this as well: Pretty much anything in excess will make you ill or kill you. Couple that with an addiction prescribed to you as the result of an early trauma, and some people may experience negative effects, anxiety or have difficulty realising the benefits of plant substances such as marijuana or 'magic mushrooms'. However, I'm sure that if they knew how to work with them, and were ready to do so, then they could cure a plethora of psychologically related conditions. There's much documented evidence to suggest that this is the case.

It is clear that teachers are *trying* to teach all of the things essential for early development, but I can also see the bigger picture, in as much as the bastardised global schooling system, as described quite succinctly in the short video given below:

RSA Animate – Sir Ken Robinson – Changing Paradigms (youtu.be/zDZFcDGpL4U)

A Theory of Everything

The overlying primary purpose of mainstream schooling appears to be to instil **an unquestionable faith in authority**, to remove the gift of inspiration, to belittle intuition, thinking or reasoning, and to teach just enough skills and knowledge to churn out a majority of mind-numbed, gullible, institutionalised, brainwashed, conformist 'workers' into the population (at every level). This is seen as being essential for maintaining the status quo, relative to the financial institution under which we live, and are governed by.

"A child educated only at school is an uneducated child"
George Santayana

"Children must be taught how to think, not what to think"
The Idealist

"Re-examine all you have been told. Dismiss what insults your soul" Walt Whitman

"What does school really teach you? Truth comes from authority. Intelligence is the ability to remember and repeat. Accurate memory and repetition are rewarded. Non-compliance is punished. Conform: Intellectually and socially" Unknown

"Education is not the learning of the facts, but the training of the mind to think" Albert Einstein

"When it comes down to it, the only knowledge that really matters is how to purify water, how to grow your own food, how to cook, how to build, and how to love. And funnily enough, we're not taught any of it in school... It's almost as if they want your head filled with crap that will never benefit your life, so you are always dependent on the government and corporations" Unknown

A Theory of Everything

"The school system is designed to teach obedience and conformity and prevent the child's natural capacities from developing" Noam Chomsky

"Formal education will make you a living; self-education will make you a fortune" Jim Rohn

---

Your mother and I spent a great deal of time investigating, and trying to get you both into Steiner Schools, but we found that under the circumstances, we were unable to orchestrate a situation where we could do it. We didn't try hard enough. I have no regrets, and I know that you're going to be just fine, especially if any of what I'm telling you now resonates with you or sinks in.

ADHD: In my opinion (and it's only my opinion), this is a term attributed to a person who has the preprogrammed propensity not to tolerate being forced to engage in anything that doesn't resonate with their higher being, or in which they are simply not interested.

Please see **Dolores Cannon** and her many astounding and enlightening dialogues on her experience as a hypnotherapist practising somnambulistic hypnosis, and with past life regression.

Documentary Reference: Dolores Cannon Presents Moving into the New Earth (youtu.be/6ucn3Xswv4Q)
Documentary Reference: Dolores Cannon My Incredible Conversation with Nostradamus (youtube.com/watch?v=LrzCnN87b34)

Dolores Cannon proposes that these people are born naturally vibrating at a higher frequency, in order to perform a specific function in assisting the human race on the earth at this point in its existence. For example, Alex has already demonstrated to us that he can generate energy in his hands at will (in the form of heat), and I believe that if he wanted to, he could easily learn, and be very effective in the practice of energy healing such as Reiki, later in life.

Alex, don't ever be overwhelmed by the label of ADHD. You are very bright, and very intelligent, and although you will have to learn how to function in society, the same as I had to, more than most kids, you simply don't tolerate anything that appears to be wasting your time, or anything that seems to be of no value to you, or that you find plain boring. The label ADHD will become quite irrelevant once you leave the education system, unless you choose to retain its influences.

You are not a bad person, and I hope you'll learn that being aware of your behaviour and how it affects others will help you to change the way you look at people, and will improve the relationships you have, in time. I know your intention is not to hurt or upset your brother or anyone else. For now, I'd like to tell you that I think you have every reason to be confident in everything you want to do, achieve or become.

Reference: 20 Life Skills Not Taught In School (tinyurl.com/yaz83zbj)
Reference: 50 Things I Wish I'd Been Taught in High School (tinyurl.com/y9dw4vdq)
Reference: The 10 Crucial Skills They Won't Teach You At School (And How To Learn Them Anyway) (tinyurl.com/ydz3nxh5)
Reference: 15 Life Skills They Don't Teach Our Kids in School (tinyurl.com/ycyw3fr6)

Appendix 4
Hacks and Quotes

List of Neat Life Hacks

NEAT LIFE HACKS: CONSIDER THAT THE ONLY DIFFERENCE BETWEEN 'SUCCESSFUL' PEOPLE AND OTHER PEOPLE IS THEIR **MENTALITY**, THE WAY THEY THINK, AND THE CONFIDENCE, DRIVE AND DETERMINATION THAT THEY HAVE IN THEMSELVES. I BELIEVE THAT ANYBODY CAN BE AS SUCCESSFUL AS THEM (ON THEIR CHOSEN PATH), IF THEY REALLY WANT TO, AND IF THEY SOUGHT THE TOOLS TO DEVELOP THEMSELVES TO BE THE WAY THEY NEED TO BE TO SUCCEED

NEAT LIFE HACKS: JUST BECAUSE SOCIETY TEACHES US TO VIEW QUOTES AND SAYINGS AS 'OLD WIVES' TALES' OR 'CLICHÉS', AND ENCOURAGES US TO PAY LITTLE ATTENTION TO THEM, IT DOESN'T MAKE THEM ANY LESS PROFOUNDLY TRUE OR BENEFICIAL TO THOSE WHO ARE BRAVE ENOUGH TO INDULGE THEM

NEAT LIFE HACKS: THEY SAY THAT A DREAMER IS A FOOL. I SAY THAT IF YOU DON'T FOLLOW YOUR DREAMS, THEN YOU ARE BY FAR THE GREATER FOOL

A Theory of Everything

NEAT LIFE HACKS: IF IT'S WHAT YOU ENJOY DOING, THEN IN GENERAL AND WITHIN REASON, IT'S PROBABLY WHAT YOU SHOULD BE DOING, AND WILL INEVITABLY LEAD TO YOUR SUCCESS AND HAPPINESS, AND THE HAPPINESS OF OTHERS AROUND YOU, IN THE LONG TERM

NEAT LIFE HACKS: OFTEN WHEN FACING CHALLENGING SITUATIONS, THE BEST PATH THAT YOU CAN FOLLOW IS GUIDED BY YOUR 'INTUITION' AND 'GUT FEELING'. I BELIEVE THAT THIS IS ONE OF THE WAYS THAT YOUR SOUL, AND INDEED THE UNIVERSE ITSELF, IS ABLE TO COMMUNICATE WITH YOU, AND GUIDE YOU ON YOUR JOURNEY.

NEAT LIFE HACKS: WHEN CHALLENGED BY A SITUATION OR TASK THAT'S WAY TOO BIG TO CONTEMPLATE ALL AT ONCE, THEN BREAK IT DOWN INTO SMALL STEPS, OR 'CHUNKS OF ELEPHANT', AS I WAS ONCE TAUGHT. WHEN YOU BREAK A TASK OR GOAL DOWN INTO 'BITE-SIZED' CHUNKS, AND TAKE THINGS STEP BY STEP, IT NOT ONLY HELPS YOU TO AVOID THE FEAR OF NOT BEING ABLE TO COMPLETE WHAT ON THE FACE OF IT SEEMS LIKE AN IMPOSSIBLE TASK, BUT ALSO MAKES THE WHOLE PROCESS IN GENERAL FAR MORE SIMPLE, AND FAR MORE HARMONIOUSLY ADAPTABLE

NEAT LIFE HACKS: DON'T WORRY IF IT DOESN'T ALL MAKE SENSE TO YOU RIGHT NOW, OR IF YOU FIND IT CONFUSING OR CHALLENGING, OR EVEN SCARY, OR DON'T EVEN WANT TO KNOW ABOUT IT, OR THINK ABOUT IT AS A RESULT. WHEN YOU'RE READY TO THINK ABOUT IT, IF YOU EVER ARE, YOU WILL DO YOUR OWN RESEARCH, AND ONE DAY IT WILL ALL JUST CLICK

NEAT LIFE HACKS: YOU'VE REALLY GOT NOTHING TO LOSE, AND EVERYTHING TO GAIN FROM TRYING TO ENJOY BEING AT SCHOOL, OR WORKING TO GAIN THE NECESSARY AND APPROPRIATE LIFE SKILLS AND EXPERIENCE

A Theory of Everything

NEAT LIFE HACKS: ACCEPT THE FACT THAT YOU UNDERSTAND THAT THE PRIMARY PURPOSE OF SCHOOL IS TO TEACH YOU TO CONFORM, AND TO ERODE YOUR ABILITY TO THINK CRITICALLY, AND TO BRAINWASH YOU TO FIT INTO OUR BASTARDISED SOCIETY. ACCEPT THIS, BUT WITH THIS KNOWLEDGE IN MIND, PAY ATTENTION TO WHAT'S IMPORTANT TO YOU, AND WHAT INTERESTS YOU, AND PRETEND VERY HARD THAT YOU ENJOY ANYTHING ELSE, NO MATTER HOW TEDIOUS YOU FIND IT, PARTICULARLY PAYING ATTENTION NOT TO DISTURBING THE 'LEARNING' OF OTHER STUDENTS. YOUR TEACHERS IN GENERAL ARE TRYING TO HELP YOU, AND WANT TO HELP YOU IN THE BEST WAY THAT THEY CAN WITHIN THE SYSTEM THEY HAVE TO WORK WITH. TRY TO MAKE IT EASIER FOR THEM, AND ABOVE EVERYTHING, TRY TO MAKE WHAT YOU DO ENJOYABLE AND FUN FOR YOURSELF (WITHOUT DISRUPTING THE REST OF THE CLASS!)

NEAT LIFE HACKS: UNDERSTAND THAT THE PRIMARY PURPOSE OF THE MEDIA, NEWS AND TELEVISION IS TO PROGRAMME OUR MINDS AND INSTITUTIONALISE US INTO BECOMING THE SLAVES THAT, FOR THE MAJORITY, WE ARE. FREE YOURSELF AND YOUR MIND BY QUESTIONING EVERYTHING YOU READ OR HEAR, PARTICULARLY IF YOU RECOGNISE THE MESSAGE AS BEING NEGATIVE, AND REJECT ANYTHING THAT DOESN'T FEEL GOOD, OR RESONATE WITH YOUR CORE. TAP INTO YOUR GUT FEELING AND INTUITION WHENEVER YOU CAN

NEAT LIFE HACKS: ALWAYS RESEARCH ANYTHING THAT YOU WANT TO KNOW ANYTHING ABOUT, PARTICULARLY THINGS THAT YOU'RE INTERESTED IN AND NEVER TELL ANYONE ANYTHING WITH CONFIDENCE, IF YOU ARE NOT HONESTLY CONFIDENT IN WHAT YOU ARE SAYING YOURSELF — PEOPLE IMPORTANT TO YOUR LIFE AND DEVELOPMENT WILL RESPECT YOU FAR MORE THIS WAY, AND WILL BE MORE INCLINED TO LISTEN WITH INTEREST TO ANYTHING YOU DO HAVE TO SAY. GENERALLY, I'D ADVISE THAT IF YOU DON'T HAVE ANYTHING INTERESTING OR INFORMATIVE TO SAY, THEN DON'T SAY IT (WHICH IS RICH COMING FROM ME, HAVING BEEN A PART-TIME GOBSHITE FOR MOST OF MY LIFE)!

A Theory of Everything

NEAT LIFE HACKS: IT ALL BEGINS WITH BELLIGERENT AND BRUTAL SELF-HONESTY IN BECOMING AWARE OF YOUR BEHAVIOUR, AND ALLOWING IT AND ACCEPTING IT, SO TO LAY THE FOUNDATIONS FOR THE TASK OF THE REBUILDING OF THE TRUE YOU, AND YOUR ULTIMATE SUCCESS

NEAT LIFE HACKS: TAKE RESPONSIBILITY FOR YOUR OWN SELF-EDUCATION AND LEARNING, AND ACTIVELY RESEARCH THE THINGS THAT INTEREST YOU AND EXCITE YOU, OR THINGS THAT YOU NEED TO UNDERSTAND BETTER TO COMPLETE A PARTICULAR TASK OR GOAL, FOR WHATEVER REASON. MAKE BEST USE OF YOUR OWN TIME TO DO THIS. YOU CAN ONLY BENEFIT FROM THIS YOURSELF IN THE LONG TERM, EVEN IF YOU THINK THAT YOU ARE DOING IT TO BENEFIT SOMEONE ELSE IN THE SHORT TERM

NEAT LIFE HACK: ALWAYS STRIVE TO PUT YOURSELF OUTSIDE OF YOUR COMFORT ZONE, WITHIN REASON. WHAT YOU'LL FIND WHEN YOU PUT YOURSELF OUTSIDE OF YOUR COMFORT ZONE IS THAT WHEN YOU LOOK BACK ON WHAT YOU'VE ACCOMPLISHED, YOU CAN SEE HOW MUCH EASIER IT LOOKS FROM THE OTHER SIDE, AND THAT THE FEAR THAT YOU EXPERIENCED PRIOR TO IT HAS EVAPORATED AND BECOME INSIGNIFICANT

NEAT LIFE HACKS: BE VERY CAREFUL WHAT YOU WISH FOR!

NEAT LIFE HACKS: MONEY IS REPORTED TO BE A MADE-UP CONCEPT USED TO CONTROL THE PEOPLE OF THE EARTH. IF MONEY IS WHAT YOU WANT, AND WHAT YOU FOCUS YOUR DESIRE ON, THEN I'M PRETTY SURE IT WILL ONLY LEAD YOU TO MISERY AND DISAPPOINTMENT. I SEE MONEY AS BEING SIMPLY A TOOL TO GET WHAT YOU WANT (BOB PROCTOR), AND TO SATISFY YOUR DESIRES. SO, IF I FOCUS WITH PASSION AND DESIRE ON WHAT IT IS I ACTUALLY WANT, AND NOT ON THE MONEY, AND IF I HAVE BELIEF AND FAITH IN THE PROCESS (WHICH I ABSOLUTELY DO), THEN THE MONEY JUST HAPPENS, AND FACILITATES WHATEVER IT IS!

A Theory of Everything

NEAT LIFE HACKS: WHAT I HAVE FOUND CONSISTENTLY IN MY LIFE IS THAT WHEN YOU REALLY WANT TO ACHIEVE, OR DO, OR GET SOMETHING, AND YOU FOCUS YOUR DESIRES ON IT, WITH BELIEF AND FAITH, FEELING AND VISUALISATION, LIFE WILL OFTEN THROW A WHOLE TON OF CONFUSION AND CHALLENGES AT YOU, JUST BEFORE THE POINT OF BREAKTHROUGH AND ACHIEVEMENT. HENCE THE EXPRESSION 'THREE FEET FROM GOLD' (TAKEN FROM THINK AND GROW RICH). WHAT I'VE FOUND IS, IT'S LIKE LIFE IS SAYING, 'ARE YOU SURE YOU REALLY WANT THIS?' AND IF YOU REMAIN STRONG AND FOCUSSED, AND PLOUGH THROUGH THE CHALLENGES, THEN IT WILL SEE THAT YOU ARE SERIOUS, AND IT WILL PROVIDE IN ABUNDANCE! ALSO, 'THREE FEET FROM GOLD' IS A GOOD 'WHAT IF, UP' STATEMENT (SEE 6.7)!

NEAT LIFE HACKS: THERE IS NO SUCH THING AS COINCIDENCE. EVERYTHING HAPPENS FOR A REASON, AND MORE OFTEN THAN NOT, BECAUSE YOU, OR SOMEONE ELSE HAS ATTRACTED IT INTO BEING OR HAPPENING. FOR THE SAME REASON, I DON'T BELIEVE IN 'LUCK' AS WE ARE LED TO UNDERSTAND IT. IT IS MY OPINION THAT FOR THE MAJORITY OF CASES, YOU CREATE YOUR OWN 'LUCK', AND I OFTEN WISH PEOPLE THE BEST OF 'TUCK'!

NEAT LIFE HACKS: DON'T EVER TAKE ANYTHING OR ANYONE FOR GRANTED. NOW THAT'S CHALLENGING, BECAUSE WE LIVE IN A SOCIETY THAT HAS TAUGHT US TO TAKE PRETTY MUCH EVERYTHING FOR GRANTED. EQUALLY, DON'T RELY ON YOUR EXPECTATIONS. IN FACT, THE BEST POSITION IS: DO NOT HAVE ANY EXPECTATIONS, BUT BE AWARE OF OPPORTUNITY AND POSSIBILITY NEVERTHELESS. BE ADAPTIVE AND INNOVATIVE TO CHANGE, BECAUSE THAT'S THE ONLY CERTAIN THING IN LIFE (OTHER THAN DEATH) – CHANGE IS INEVITABLE, AND THOSE WHO RESIST IT WILL SUFFER IN THE LONG TERM

A Theory of Everything

NEAT LIFE HACKS: I FIND GREAT PLEASURE IN DOING THINGS FOR OTHERS AND HELPING OTHERS. SOMETHING THAT I LEARNED IS THAT WHAT GOES AROUND, COMES AROUND. WHAT I FOUND IS CRITICALLY IMPORTANT TO UNDERSTAND ABOUT THIS STATEMENT IS THAT GIVING MUST BE UNCONDITIONAL, AND ONE MUST NEVER EXPECT TO BE GIVEN BACK DIRECTLY AS A RESULT, OR TO EXPECT ANYTHING IN RETURN DIRECTLY FROM THOSE THAT YOU GIVE TO. YOU SEE, THE WAY IT WORKS IS THAT IF YOU GIVE (OR PAY IT FORWARD), YOU WILL GET YOUR KINDNESS OR ENERGY OR LOVE BACK, BUT IT MAY COME FROM A DIFFERENT PERSON, ANGLE OR SOURCE

NEAT LIFE HACK: DON'T BE AFRAID TO WISH FOR WHAT YOU REALLY WANT, AND NOT JUST WHAT SOCIETY LEADS YOU TO PERCEIVE AS BEING SOCIALLY 'REASONABLE' OR 'REALISTIC'

NEAT LIFE HACKS: WHEN CONFRONTED WITH A SITUATION WHERE YOU HAVE TO MAKE A DECISION, HAVE FAITH IN YOUR INTUITION AND GO WITH THE FIRST THING THAT SPRINGS TO MIND, OR THE THING YOU FEEL MOST STRONGLY TOWARDS, EVEN IF IT SEEMS COUNTER-INTUITIVE. MORE OFTEN THAN NOT YOU WILL HAVE MADE THE RIGHT DECISION; HOWEVER, IF YOU SUBSEQUENTLY FEEL AS THOUGH YOU COULD HAVE MADE A BETTER DECISION, BE GRATEFUL FOR THE LESSON THAT YOU LEARNED, AND APPLY THE GAINED KNOWLEDGE IN THE FUTURE

NEAT LIFE HACKS: IF YOU'RE PRACTISING THE LAW OF ATTRACTION AND AFFIRMATION, THEN THERE ARE A NUMBER OF KEY INGREDIENTS THAT YOU MUST EMPLOY, AND ONE OF THOSE IS ACTION! YOU CAN'T DO AN AFFIRMATION AND ASK FOR SOMETHING, AND THEN SIT ON YOUR ARSE DOING NOTHING WAITING FOR IT TO HAPPEN! IT'S NOT GOING TO HAPPEN THAT WAY! YOU MUST TAKE SOME ACTIONS AND STEPS TOWARDS YOUR GOALS, NO MATTER WHAT THEY ARE

A Theory of Everything

NEAT LIFE HACK: IF YOU EVER FIND YOURSELF IN A STATE OF FIGHT-OR-FLIGHT, AND YOU WANT TO LEVERAGE THE FIGHT, RATHER THAN THE FLIGHT, THEN SIMPLY TAKE THREE DEEP BREATHS BEFORE YOU ENGAGE, AND THINK POSITIVE THOUGHTS ABOUT HOW YOU SEE YOURSELF ACHIEVING WHAT YOU WANT TO DO. FEEL HOW HAPPY IT MAKES YOU

NEAT LIFE HACK: I FIND IT EXTRAORDINARY THAT ONE OF THE MOST FUNDAMENTAL MECHANISMS OF LIVING, BREATHING, IS TAKEN FOR GRANTED AND ALMOST IGNORED AND MISUNDERSTOOD BY MANY OF THE POPULATION, WHO SEEM TO BE PRONE TO SHALLOW BREATHING FOR THE MAJORITY OF THE TIME. BREATHE DEEPLY WHENEVER YOU CAN, AND ALLOW YOUR BODY TO ABSORB THE OXYGEN IT RECEIVES

NEAT LIFE HACK: YES, YOU REALLY CAN CHOOSE HOW TO THINK AND SUBSEQUENTLY FEEL ABOUT A SITUATION. IT TAKES PRACTICE, FOR SURE, BUT IN MY OPINION, THERE IS NO VALUE TO ANYONE IN CHOOSING TO FEEL RESENTFUL, ANGRY, DISAPPOINTED, BITTER, REVENGEFUL, JEALOUS, OR ANY OF THE OTHER NEGATIVE EMOTIONS THAT YOU WOULD OTHERWISE LET YOURSELF EXPERIENCE. TRUST THAT THERE IS POSITIVE TO BE FOUND IN ANY EXPERIENCE YOU ENDURE, AND ACTIVELY FIND THE POSITIVE

NEAT LIFE HACKS: AT A FUNDAMENTAL LEVEL YOU MUST BE 'SELFISH' WITH YOURSELF INITIALLY, IN ORDER TO BE ABLE TO ENJOY THE BEAUTY AND WONDER OF TRULY GIVING TO OTHERS, CARING FOR OTHERS AND LOVING OTHERS IN THE FUTURE. LOVE YOURSELF BEFORE YOU LOVE ANYTHING OR ANYONE ELSE, AND THAT LOVE WILL RADIATE HARMONIOUSLY OUTWARDS

A Theory of Everything

NEAT LIFE HACK: WHEN COMMUNICATING WITH OTHERS, IN PARTICULAR
THOSE YOU RESPECT, ASK YOURSELF WHETHER YOU ARE SURE THAT WHAT YOU
ARE SAYING IS TRUE, OR WHETHER YOU ARE SIMPLY REGURGITATING HEARSAY OR
RUMOUR, AND ADJUST YOUR DIALOGUE ACCORDINGLY. IF YOU DO THIS, THEN
THERE IS A FAR GREATER CHANCE THAT YOU WILL BE TRUSTED AND TREATED
WITH RESPECT IN GENERAL, AND THAT THE RECEIVER IS MORE LIKELY TO BE MORE
HONEST WITH YOU AS A RESULT

NEAT LIFE HACK: TRUST EVERYONE WITHIN REASON IMMEDIATELY UNTIL SUCH A
POINT THAT THEY LET YOU DOWN, AND IF THIS HAPPENS THEN LET THEM DOWN
GENTLY AND REMOVE THEM FROM YOUR LIFE

NEAT LIFE HACKS: IF YOU EXPERIENCE A NEGATIVE THOUGHT (STEMMING FROM
A SUBCONSCIOUS BELIEF, OR MEMORY AND ASSOCIATED FEELING OR EMOTION)
AT ANY TIME, YOU WILL FIND THAT IF YOU CONSCIOUSLY ACKNOWLEDGE IT,
ACCEPT IT, AND ALLOW IT, THEN IT HAS NO PLACE TO GO, AND IT WILL SIMPLY
DISAPPEAR. IT'S A BIT LIKE WHEN SOMEONE IS ANNOYING YOU, ANTAGONISING
YOU, OR WINDING YOU UP AND YOU JUST ACCEPT THAT THEY ARE THERE BUT
DON'T TAKE THEIR BAIT, AND DON'T RISE TO THE SITUATION, THEY WILL GET
BORED AND EVENTUALLY LEAVE YOU ALONE

NEAT LIFE HACKS: IF A NEGATIVE THOUGHT COMES INTO YOUR MIND AND YOU
WANT TO ERADICATE IT, THEN SIMPLY REPEAT THE WORDS 'I LOVE YOU' IN YOUR
MIND, UNTIL THE NEGATIVE THOUGHT HAS VANISHED

NEAT LIFE HACKS: THE WHAT IF 'UP' TECHNIQUE. THIS IS A VERY NEAT AND
SIMPLE TECHNIQUE TO HELP YOU WITH NEGATIVE THOUGHTS AND RESULTING
EMOTIONS. FOR EXAMPLE, CATCH YOURSELF THINKING 'WHAT IF THE WEATHER
IS GOING TO BE TERRIBLE FOR THE MATCH TOMORROW?' AND CHANGE
THAT THOUGHT TO 'WHAT IF THE WEATHER IS GOING TO BE A SCORCHER

TOMORROW, AND WE HAVE A LOVELY DAY?', OR 'WHAT IF, WHATEVER THE WEATHER DOES, I'M GOING TO BE HAPPY ANYWAY'. IT'S FREE, AND YOU'VE GOT NOTHING TO LOSE, AND IF IT DOESN'T MAKE YOU INSTANTLY HAPPIER, THEN I'LL GIVE YOU YOUR MONEY BACK!

NEAT LIFE HACK: WHEN YOU FEEL THAT YOU ARE THREATENED BY NEGATIVITY OR NEGATIVE ENERGY, ASK TO BE CONNECTED TO THE SOURCE, THEN ASK YOUR GUARDIAN ANGEL TO PUT A SIX-FOOT FORCE FIELD AROUND YOU TO PROTECT YOU, AND VISUALISE THAT BARRIER GOING UP AROUND YOU.

NEAT LIFE HACKS: JUST THINKING ABOUT WHAT OTHERS MIGHT BE SAYING OR THINKING ABOUT YOU IN A NEGATIVE RESPECT IS NEGATIVE SELF-THINKING, AND IF YOU'RE NOT CAREFUL, IN THIS WAY CAN BECOME A SELF-FULFILLING PROPHECY

NEAT LIFE HACKS: WHENEVER YOU FEEL CHALLENGED GOING INTO A SITUATION OR ACTIVITY, TAKE THE TIME TO SET YOUR INTENTIONS FIRST, AND FOCUS ON EXPECTING A POSITIVE OUTCOME. INTENTION IS A VERY POWERFUL THOUGHT-PROCESS, WHICH CAN BE USED EFFECTIVELY AT ANY TIME THAT YOU WANT TO HAVE A POSITIVE INFLUENCE ON A FUTURE EVENT. IN FACT, I'D SAY, WHERE POSSIBLE, ALWAYS TRY TO SET YOUR INTENTION BEFORE DOING ANYTHING

NEAT LIFE HACKS: IF YOU TAKE THE TIME TO ANALYSE THE LANGUAGE THAT YOU USE, RELATIVE TO THE FEELINGS THAT YOU'RE EXPERIENCING, AS A RESULT, YOU CAN BEGIN TO CHANGE THE LANGUAGE THAT YOU USE TO BE MORE POSITIVE, AND SUBSEQUENTLY RAISE YOUR VIBRATIONAL FREQUENCY. IN DOING SO, YOU WILL INEVITABLY FEEL BETTER, MORE POSITIVE AND HAPPY, AND YOU WILL IN TURN ATTRACT MORE POSITIVES INTO YOUR LIFE

A Theory of Everything

NEAT LIFE HACKS: IF YOU'RE NOT PUTTING YOURSELF OUT THERE AND MAKING MISTAKES, AND LEARNING FROM THEM, THEN INEVITABLY YOU'RE NOT MOVING FORWARD IN LIFE

NEAT LIFE HACKS: REMEMBER THAT WHEN YOU APOLOGISE TO SOMEONE, OR SAY 'SORRY', THEN THE MOST POSITIVE POSITION YOU CAN PUT YOURSELF IN IS TO USE THAT EXPERIENCE TO ADDRESS YOUR BEHAVIOUR IN THE FUTURE TO ENSURE THAT YOU DO NOT MAKE THE SAME MISTAKE AGAIN

NEAT LIFE HACKS: DON'T EVER BE AFRAID TO SAY THAT YOU DON'T UNDERSTAND SOMETHING, OR BE AFRAID OF ASKING 'STUPID' QUESTIONS. IN YOUR QUEST FOR UNDERSTANDING, THERE ARE NO STUPID QUESTIONS, AND, ACTUALLY, SOMETIMES THE MOST STUPID QUESTIONS ARE THE MOST THOUGHT-PROVOKING AND PROFOUND

NEAT LIFE HACKS: RECOGNISE AND ACCEPT THAT EVERYONE ON EARTH IS A UNIQUE AND VALUABLE INDIVIDUAL, AND THAT PREJUDICE AND CLASS STATUS, FOR EXAMPLE, ARE A SOCIALLY IMPOSED POISON OF THE MIND. TREAT EVERYBODY THE SAME, NO MATTER WHO THEY ARE, WHERE THEY'RE FROM, WHAT THEY DO, WHAT THEIR LEVEL OF WISDOM OR INTELLIGENCE IS, OR WHAT COLOUR THEIR SKIN IS, BECAUSE THEY ARE SOULS ON THE SAME JOURNEY AS YOU ARE, JUST AT A DIFFERENT STAGE, AND WITH DIFFERENT PURPOSE

A Theory of Everything

NEAT LIFE HACKS: AS A GENERAL RULE OF THUMB, I'D RECOMMEND THAT IF YOU DON'T HAVE ANYTHING GOOD TO SAY ABOUT ANYONE OR ANYTHING, THEN SIMPLY SAY NOTHING AT ALL. AND ONCE YOU'VE RECOGNISED THAT YOU DON'T HAVE ANYTHING GOOD TO SAY ABOUT SOMETHING OR SOMEONE, I'D RECOMMEND TRYING TO LOOK FOR SOMETHING GOOD TO THINK ABOUT THEM OR IT — NO MATTER HOW CHALLENGING THAT MAY BE. EQUALLY, TAKE A GOOD LOOK AT THE NEGATIVE YOU HAVE RECOGNISED IN THEM, AND SEE IF YOU CAN IDENTIFY THAT TRAIT WITHIN YOURSELF

NEAT LIFE HACKS: IF YOU CATCH YOURSELF COMPLAINING ABOUT SOMETHING, THEN TRY TO CHANGE THE WAY YOU THINK ABOUT THE SITUATION TO BE POSITIVE. IT DOESN'T MATTER WHAT THE POSITIVE THINKING IS, IT JUST NEEDS TO BE POSITIVE, AND I AM SURE THAT IN DOING THIS, IF NOTHING ELSE, IT WILL MAKE YOU FEEL BETTER ABOUT THE SITUATION. AS A BONUS, I BELIEVE, IT WILL ALSO BRING YOU MUCH CLOSER TO GAINING WHAT YOU ACTUALLY WANT

NEAT LIFE HACK: DON'T BEAT YOURSELF UP FOR ANYTHING. WHAT'S DONE IS DONE. THE BEST POSITION YOU CAN PUT YOURSELF IN IS TO ACCEPT THAT YOU'VE MADE A MISTAKE, AND TO USE THE OPPORTUNITY TO ACTIVELY WORK TOWARDS LEARNING FROM IT, AND TO IMPROVE YOUR BEHAVIOUR IN THE FUTURE

NEAT LIFE HACK: REMOVE NEGATIVE PEOPLE FROM YOUR LIFE. NEGATIVITY BREEDS NEGATIVITY AND ONLY ACTS TO ERODE YOUR POSITIVITY AND HIGHER STATE OF RESONANCE (VIBRATIONAL FREQUENCY). YOU CANNOT HELP THESE PEOPLE. TRY TO SURROUND YOURSELF WITH PEOPLE WHO INSPIRE YOU AND BUILD YOUR POSITIVE ENERGY, AND VICE VERSA

A Theory of Everything

NEAT LIFE HACKS: BE AWARE OF THE FACT THAT WHEN YOU FIRST GET INTO A RELATIONSHIP, IT IS LIKELY THAT BOTH YOU AND YOUR PARTNER WILL INITIALLY TELL EACH OTHER ABOUT WHO YOU VIEW YOURSELVES TO BE IN YOUR CONSCIOUS MINDS, AND THAT THIS MAY BE A VERY DIFFERENT REALITY TO WHO YOU ARE AND HOW YOU ACTUALLY BEHAVE, BASED ON THE BELIEFS YOU HOLD IN YOUR SUBCONSCIOUS MINDS. I CONSIDER THAT THE BEST RELATIONSHIPS WORK WHEN YOU ARE TRULY BEST FRIENDS BEFORE YOU ARE LOVERS, OR COMMITTED TO THE RELATIONSHIP

NEAT LIFE HACKS: TO PUT YOURSELF IN A BETTER POSITION TO GIVE AND RECEIVE LOVE UNCONDITIONALLY IN A RELATIONSHIP, YOU MUST FIRST TRULY LOVE YOURSELF. YOU MUST PUT YOURSELF IN A POSITION WHEREBY YOU HAVE NO EXPECTATIONS OR DEMANDS. YOU MUST ACCEPT AND MAKE ALLOWANCE FOR WHO YOUR PARTNER TRULY IS, AND YOU MUST HOLD NO GOVERNANCE OVER THEM

NEAT LIFE HACKS: REMEMBER THAT WHEN ANYTHING IS PRESENTED AS A THEORY IT IS REFERRED TO IT AS SUCH BECAUSE THAT'S EXACTLY WHAT IT IS, A THEORY! FOR EXAMPLE, THE THEORY OF RELATIVITY IS A THEORY. DARWIN'S EXPLANATION OF EVOLUTION IS A THEORY. GRAVITY IS A THEORY. PARTICLE PHYSICS AND PARTICLE GEOMETRY ARE A THEORY. "THERE ARE NO PARTICLES, JUST RESONANCE AND FREQUENCY" – SEE DOUBLE-SLIT EXPERIMENT! QUANTUM THEORY AND HELIOCENTRISM, THE THEORY THAT THE EARTH AND PLANETS REVOLVE AROUND THE SUN – NOW THAT'S A GOOD ONE! ETC...

A Theory of Everything

List of Quotes

"I once heard a wise man say there are no perfect men. Only perfect intentions." Morgan Freeman

"Never doubt that a small group of thoughtful, committed citizens can change the world. Indeed, it is the only thing that ever has" Margaret Mead

"Our alienation goes to the roots...We are born into a world where alienation awaits us. We are potentially men, but are in an alienated state...the ordinary person is a shrivelled, desiccated fragment of what a person can be. As adults, we have forgotten most of our childhood, not only its contents but its flavour; as men of the world, we hardly know of the existence of the inner world...The condition of alienation, of being asleep, of being unconscious, of being out of one's mind, is the condition of the normal man...between us and It [our true selves or soul] there is a veil which is more like fifty feet of solid concrete...The outer divorced from any illumination from the inner is in a state of darkness. We are in an age of darkness...We are all murderers and prostitutes...We are bemused and crazed creatures, strangers to our true selves, to one another' (see par. 123 of FREEDOM)." R. D. Laing

"Here's to the crazy ones, the misfits, the rebels, the troublemakers, the round pegs in the square holes... the ones who see things differently – they're not fond of rules, and they have no respect for the status quo.... You can quote them, disagree with them, glorify or vilify them, but the only thing you can't do is ignore them because they change things... They push the human race forward, and while some may see them as the crazy ones, we see genius, because the people who are crazy enough to think that they can change the world, are the ones who do." Steve Jobs

A Theory of Everything

"The beginning is near. When the earth is ravaged and the animals are dying, a new tribe of people shall come onto the earth from many colours, classes, and creeds, and by their actions shall make the earth green again. They will be known as the warriors of the rainbow" Native American

"Success is when we wake up and notice that the true purpose of life is simply being yourself" Linda Mallory

"To be old and wise, you must first be young and stupid" SunGazing

"The 3 C's of life: Choices, Chances and Changes. You must make a choice to take a chance, or your life will not change" Zig Ziglar

"I choose… To live by choice, not by chance. To be motivated, not manipulated. To be useful, not used. To make changes, not excuses. To excel, not compete. I choose self-esteem, not self-pity. I chose to listen to my inner voice, not to the random opinions of others" Unknown

"The Secret to Life… is that you have all the power. If you can change your mind, you can change your life. You don't need any amount of money, education, or connections. You just need to be in control of your mind. Program your mind to believe in yourself. Know that you have what it takes to achieve greatness, and watch greatness manifest in your life. When it comes to the law of attraction, all you're really doing is programming your mind. This is the secret that so many people overlook in life. The key to success is overcoming your limiting beliefs and aligning yourself with the energy you want in your life" Third Eye Thoughts

A Theory of Everything

"If you end up with a boring miserable life because you listened to your mom, your dad, your teacher, your priest, or some guy on television telling you how to do your shit, then you deserve it"
Frank Zappa

"Your time is limited, so don't waste time living someone else's life" Steve Jobs

"Wanting to be someone else is a waste of who you are" Kurt Cobain

"Can you remember who you were, before the world told you who you should be?" Danielle LaPorte

"If you can dream it, you can do it" Walt Disney

"Dreams at first seem impossible, then seem improbable, and finally, when we commit ourselves, become inevitable" Mahatma Gandhi

"Anything is possible" James Hunt

"The reality is that no one is actually completely average and has at least one skill or talent, be that as yet undeveloped, which is well above average. In a world that has perpetuated the cult of the average and valued sameness, conformity is losing its grip on the reins as authenticity and exceptions rule. There's one thing that you're a world champion at. No one does it better than you. You're the best at being you. And when you're being your best self, your world will transform from a round hole to the shape of your square peg" Sunil Bali

"Instinct is something that transcends knowledge" Nikola Tesla

A Theory of Everything

"Whether you believe that you can, or you believe that you can't, you're right" Henry Ford

"If you want to find the secrets of the Universe, think in terms of energy, frequency and vibration" Nikola Tesla

"The day science begins to study non-physical phenomenon, it will make more progress than it made in all the previous centuries of its existence" Nikola Tesla

"Everything is energy and that's all there is to it. Match the frequency of the reality you want and you cannot help but get that reality. It can be no other way. This is not philosophy. This is physics!" Albert Einstein

"Nothing rests. Everything moves. Everything vibrates. At the most fundamental level, the Universe and everything which comprises it is purely vibratory energy manifesting itself in different ways. The Universe has no 'solidity' as such. Matter is merely energy in a state of vibration" Unknown

"Everything we call 'real' is made of things that cannot be regarded as 'real'. Those who are not shocked when they first come across quantum physics cannot possibly have understood it" Niels Bohr

"Knowledge Protects; Ignorance Endangers"

"You can ignore reality, but you can't ignore the consequences of reality"

Aeon of Horus – The Occult History of NASA

A Theory of Everything

"For what it's worth, it's never too late, or in my case, too early to be whoever you want to be. There's no time limit, stop whenever you want. You can change or stay the same, there are no rules to this thing. We can make the best or the worst of it. I hope you make the best of it. And I hope you see things that startle you. I hope you feel things you never felt before. I hope you meet people with a different point of view. I hope you live a life you're proud of. If you find that you're not, I hope you have the courage to start all over again." F. Scott Fitzgerald

"The general population doesn't know what's happening, and it doesn't even know that it doesn't know" Noam Chomsky

"What doesn't kill you, makes you stronger"

"A child educated only at school is an uneducated child" George Santayana

"The school system is designed to teach obedience and conformity and prevent the child's natural capacities from developing" Noam Chomsky

"Education is what remains after one has forgotten what one has learned in school" Albert Einstein

"Cheers to all of those who do their own research. In the age of information, ignorance is a choice" Leonardo DiCaprio

"Our ancestors devised every method imaginable to alert us to a single fact; now is the time of the most extraordinary conditions and opportunities that accompany the rarest of events, the shift from one world age to the next" Gregg Braden 'Fractal Time'

A Theory of Everything

"Science suggests that the next step of human evolution will be marked by awareness that we are all independent cells within the super-organism called humanity" Bruce Lipton and Steve Bhaerman. 'Spontaneous Evolution'

"I believe it will be the magnetic influence produced by the sun that will usher in what is described by our ancestors as 'the transition' bringing us to a new state of being" Mitch Battros

"Never wish life were easier, wish that you were better" Jim Rohn

"If you want to find the secrets of the Universe, think in terms of energy, frequency and vibration" Nikola Tesla

"Close your eyes and let the mind expand. Let no fear of death or darkness arrest its course. Allow the mind to merge with Mind. Let it flow out upon the great curve of consciousness. Let it sour on the wings of the great bird of duration, up to the very Circle of Eternity" Hermes Trismegistus

"Being 'realistic' is the most commonly trodden road to mediocrity" Will Smith

"Life's not a spectator sport. If watching is all you're going to do, then you're going to watch life go by without you" The Hunchback of Notre Dame

"Truly, 'thoughts are things', and powerful things at that, when they are mixed with definiteness of purpose, persistence, and a burning desire for their translation into riches, or other material objects" Napoleon Hill

A Theory of Everything

"Remember to look up at the stars and not down at your feet. Try to make sense of what you see and wonder about what makes the universe exist. Be curious. And however difficult life may seem, there is always something you can do and succeed at. It matters that you don't just give up." Stephen Hawking

"You are the average of the five people you spend most of your time around" Practical Psychology

"History records that the money chargers have used every form of abuse, intrigue, deceit, and violent means possible to maintain their control over government by controlling money and its issuance" James Madison

"Meditation makes you innocent, it makes you childlike. In that state, miracles are possible. That state is pure magic. A great transformation happens – In innocence you transcend the mind, and to transcend the mind is to become the awakened one, the enlightened one" Osho

"True healing will always begin with your thoughts. Master your thoughts and you will master your life" April Peedess

"If every 8-year-old in the world is taught meditation, we will eliminate violence from the world within one generation" Dalai Lama

"Please grant me the serenity to accept the things I cannot change, the courage to change the things I can, and the wisdom to know the difference" The Serenity Prayer

"Worrying merely impairs your ability to maximise the impact you have on situations at hand – chill out!" Mr. Jonathan R. Tuck (aged 16)

A Theory of Everything

"As you begin to be grateful for what others take for granted, that vibration of gratitude makes you more receptive to good in your life" Michael Bernard Beckwith

"It is through gratitude for the present moment that the spiritual dimension of life opens up" Eckhart Tolle

"Gratitude, like faith, is a muscle. The more you use it, the stronger it becomes, and the more power you have to use it on your behalf. If you do not practice gratefulness, its benefaction will go unnoticed, and your capacity to draw on its gifts will be diminished. To be grateful is to find blessings in everything. This is the most powerful attitude to adopt, for there are blessings in everything" Alan Cohen

"You know, the ancient Egyptians had a beautiful belief about death. When their souls got to the entrance to Heaven, the guards asked two questions. Their answers determined whether they were able to enter or not. 'Have you found joy in your life?' 'Has your life brought joy to others?'" Morgan Freeman

"All you need is love" John Lennon

"I believe in the kind of love that doesn't demand me to prove my worth and sit in anxiety. I crave a natural connection, where my soul is able to recognise a feeling of home in another. Something free-flowing, something simple. Something that allows me to be me without question" Joey Palermo

"The Universe only pretends to be made of matter. Secretly, it is made of love" Daniel Pinchbeck

"Money is merely a tool to facilitate the things of your desires" Bob Proctor

A Theory of Everything

"Be <u>very</u> careful what you wish for" Dr Jonathan R. Tuck

"When I run after what I think I want, my days are a furnace of stress and anxiety; if I sit in my own place of patience, what I need flows to me, and without pain. From this I understand that what I want also wants me, is looking for me and attracting me. There is a great secret here for anyone who can grasp it" Rumi

*"There only **is**; has only ever **been**; and only ever will **be**, **right here and now"** Dr* Jonathan R. Tuck

*"Make decisions **quickly**, and change them **slowly**"* Bob Proctor

"The most dangerous psychological mistake is the projection of the shadow on to others: this is the root of almost all conflicts" Carl Jung

"You can tell more about a person by what he says about others than you can by what others say about him" Audrey Hepburn

"Being honest may not get you many friends, but it will get you the right ones" John Lennon

"I don't think I'm smarter than anyone, I just see things that others don't see. I'm not here to judge, I'm here to point out things you may or may not understand, and make you think" Keen James

"Love everyone you meet from the moment you meet them. Most people will be lovely and love you back, and you can achieve the most wonderful things. But get rid of any of the b@stards that let you down" Joanna Lumley

A Theory of Everything

"There are some people who could hear you speak a thousand words, and still not understand you. And there are others who will understand without you even speaking a word" Unknown

"Vibrational Frequency:
Raising your vibration – Gratitude, kindness, love, joy, passion, forgiveness, acceptance, sunshine, walking in nature (particularly barefooted – grounding), breathing deeply, yoga, meditation, exercise, laughing, smiling, hugging, singing, dancing, raw wholefoods, greens, fruits, nuts, creativity, relaxing music. Lowering your vibration – Junk food, alcohol, toxic relationships, negative thoughts, environmental toxins, toxic products, excess red meat, white sugar and sweets, medication, radiation, yelling, arguing, holding onto the past, anger, resentment, guilt"
Unknown

"The privilege of a lifetime is to become who you truly are" Carl Jung

"You have to look deeper, way below the anger, the hurt, the hate, the jealousy, the self-pity, way down deeper where the dreams lie. Find the dream. It's the pursuit of the dream that heals you" Lakota prayer

"Who gives a fuck?" Richard Davies

"My philosophy is: It's none of my business what people say of me and think of me. I am what I am, and I do what I do. I expect nothing and accept everything. And it makes life so much easier" Anthony Hopkins

"Being honest may not get you many friends, but it will get you the right ones" John Lennon

A Theory of Everything

"Confidence is not: 'They will like me'. Confidence is: 'I'll be fine if they don't'" Lessons Learned

"One day it just clicks. You realise what's important and what isn't. You learn to care less about what other people think about you and more about what you think of yourself. You realise how far you've come, and you remember when you thought things were such a mess that you would never recover. And you smile. You smile because you are truly proud of yourself and the person that you've fought to become" Word Porn

"If you suffer it is because of you, if you feel blissful it is because of you. Nobody else is responsible – only you, and you alone. You are your hell and your heaven, too" Osho

"Take all the money that we spend on weapons and defences each year and instead spend it feeding and clothing and educating the poor of the world, which it would pay for many times over, not one human being excluded, and we could explore space together, both inner and outer, forever, in peace" Bill Hicks

"People ask me 'Are you proud to be an American? And I say, 'I don't know, I didn't have a lot to do with it. My parents fucked there, that's about all.' I hate patriotism, I can't stand it. It's a round world last time I checked.'" Bill Hicks

"See if you can catch yourself complaining, in either speech or thought, about a situation you find yourself in, what other people do or say, your surroundings, your life situation, even the weather. To complain is always non-acceptance of what is. It invariably carries an unconscious negative charge. When you complain, you make yourself into a victim. When you speak out, you are in your power. So change the situation by taking action or by speaking out if necessary or possible; leave the situation or accept it. All else is madness." Eckhart Tolle

A Theory of Everything

"We've been infected with this idea that love is an emotion only felt between two people. But love is universal. An Energy. A contagious force. To offer money to a homeless man is to love. To smile at a stranger is to love. To be grateful, to be hopeful, to be brave, to be forgiving, to be proud, is to love" A. R. Lucas

"We live in a world where we have to hide to make love, while violence is practised in broad daylight" John Lennon

"Lightning shines in the darkness of the night, but the darkness is not part of the lightning. The only relation between the two is that lightning only stands out at night, only in the darkness. And the same is true of sex. There is a realization, an exhilaration, a light that shines in sex, but that phenomenon is not from sex itself. Although it is associated with it, it is just a by-product. The light that shines in orgasm transcends sex; it comes from beyond. If we can comprehend this experience of the beyond we can rise above sex. Otherwise, we will never be able to." Osho

"The pharmaceutical industry in general is a very powerful corporate-run body who do not care about anybody's health, wellbeing or recovery. They only care about making money and selling you their chemically synthesised drugs and treatments. They have developed these for the primary purpose of not making you better, but prolonging your illness in order to sell you more drugs, and make more money. Generally, they don't care how they do it. It is a multi-billion-dollar industry, and many of the common ailments known to man are rumoured to have been equally synthesised by these exact bodies for the purpose of making more money from the general population"
Dr Jonathan R. Tuck

A Theory of Everything

"The food industry in general is a very powerful corporate-run body controlled primarily by 8 of the top global corporations, including pharmaceutical companies, who do not care about anybody's nutritional health or wellbeing. They only care about making money from you and selling you the food that they have produced in the cheapest way possible, regardless of health or nutritional value. It is a multi-billion-dollar industry, and they really don't care about us, or the food that we eat or our resulting health" Dr Jonathan R. Tuck

"You're beginning to understand, aren't you? That the whole world is inside you: in your perspectives and in your heart. That to be able to find peace, you must be at peace with yourself first; and to truly enjoy life, you must enjoy who you are; and once you learn how to master this, you will be protected from everything that makes you feel like you can't go on, that with this gift of recognising yourself, even when you are alone, you will never be lonely." The Spiritual Compass

"I love people who make me laugh. I honestly think it's the thing I like most, to laugh. It cures a multitude of ills. It's probably the most important thing in a person." Audrey Hepburn

"We cannot solve problems by using the same kind of thinking we used when we created them" Albert Einstein

"None of us are getting out of here alive, so please stop treating yourself like an afterthought. Eat the delicious food. Walk in the sunshine. Jump in the ocean. Say the truth that you're carrying in your heart like hidden treasure. Be silly. Be kind. Be weird. There is no time for anything else" Anthony Hopkins

A Theory of Everything

"Accept – then act. Whatever the present moment contains, accept it as if you had chosen it. Always work with it, not against it… This will transform your whole life." Eckhart Tolle

"We learn plenty about our physical health, but what about our emotional and mental health? What about our inner worlds? Could there be any topic more relevant to students and young adults than understanding and managing their stress, anxiety, and emotions? If mindfulness and emotional awareness was as essential to the public school curriculum as Common Core math strategies, we just might raise a healthier generation of humans"
Babble.com

"Whether you leave the country or just leave your town, try to plan at least one trip every six months. Use websites like Student Universe (tinyurl.com/2pndxz) to find affordable deals, and communicate with your university about travel grants. Consider going on a gap year. According to an interview (tinyurl.com/ ybu59bzc) with Dale Stephens, the founder of Gap Year (uncollege.com/program), students that take a gap year have higher graduation rates and GPAs than students who opt not to take one" UnCollege.Org

"Be that impressive 20-something who knows the ins and outs of their bank account, their credit score, and their investments. Create an account at Mint.com (www.mint.com/) and start a realistic budget for yourself. Set aside 10% of every pay check you get, and look out for free MOOCs about personal finance (tinyurl.com/y7ojsk4q). Get familiar with helpful money-saving blogs like 20somethingfinance (tinyurl.com/5o42ec), The Billfold (thebillfold.com/) and The Financial Diet (thefinancialdiet.com/)"
UnCollege.Org

A Theory of Everything

"Remember that time spent on your relationships is not time wasted, even if those relationships eventually end. Every friendship and relationship you form can teach you how to strike the right balance in a life partnership. Write down the qualities you are looking for in a partner, and focus on the qualities you have to offer. Make sure to protect time each week to spend on your relationships: don't let yourself become a one-sided person"
UnCollege.Org

"We take classes and a test before getting a driver's license. We take Lord knows how many exams before getting into college. We're even offered a variety of parenting/birthing/ breastfeeding classes before having a baby. And yet I could walk into a courthouse with a simple registration and some makeshift rings and call it a marriage. How can something so complicated and important – something that affects everything from our money to our health to our happiness – have next to no training or instructions? This is another thing that should be learned at home in theory, except many kids have really crappy relationship role models because their parents had crappy role models because THERE'S NO EDUCATION ON MAINTAINING RELATIONSHIPS" Babble.com

"Take time to determine what puts you in 'flow.' (tinyurl.com/ myu7k28) Take classes outside of your subject, and think creatively about ways that you can make a living. And be sure to check out the UnCollege blog (tinyurl.com/y95eowvd) for more tips about pursuing your passions and building a singular, extraordinary life" UnCollege.Org

"Yes, I'm a conspiracy nut, because it's better than being a brainwashed sheep" Jack Nicholson

A Theory of Everything

"Scientists are now realizing that the newest crop of humans have an unprecedented ability to multitask, probably due to neuroplasticity (our brains' ability to adapt and change to the environment). New York magazine reported that kids can '[conduct] 34 conversations simultaneously across six different media, or pay attention to switching between attentional targets in a way that's been considered impossible.' But with the give comes the take, and studies show that these kids have less of an attention span than ever before. Perhaps the best thing we can teach these kids is to single-task, and to really listen and focus, rather than succumb to every distraction like a dog in a field of squirrels" Babble.com

"What if I told you that television is the monster in your home, and it's called a program for a reason. Your television is nothing more than an electronic mind-altering device that has been designed to psychologically change the way you view reality" Morgan Freeman

"I've missed more than 9000 shots in my career. I've lost almost 300 games. 26 times, I've been trusted to take the game winning shot and missed. I've failed over and over and over again in my life. And that is why I succeed" Michael Jordan

"You have to fail early, you have to fail often, and you have to fail forward" Will Smith

"That's how you become a great man. Hang your balls out there" Jerry Maguire

"A foolish faith in authority is the worst enemy of the truth" Albert Einstein

A Theory of Everything

"Start practising meditation for just ten minutes a day. Take advantage of on-campus counselling options. Set boundaries for when you allow yourself to check your inbox, and try to limit your 'screen time' in the mornings and evenings. Know how to ask for a 'personal day'. Professors and employers will respect you for knowing your boundaries" UnCollege.Org

"Read the biographies of people you admire, and make a note of the ways in which they coped with set-backs. Always remind yourself of the big picture, and learn not to sweat the small stuff (dontsweat.com/" UnCollege.Org

"Try one-on-one networking with older students, professors, and others in your field. Don't be afraid to reach out with an unsolicited email. Remember, they were once in your position! And remember that serving as a connector – being able to link two friends together – is just as important as forging connections for yourself" UnCollege.Org

"Start by making yourself a person who doesn't second-guess the choices you make, even small ones. Have to choose which restaurant to check out for dinner? Go with your first instinct. Don't know which class to sign up for? Go with the one that 'feels right.' And be sure to pay attention to when something feels 'off': trust your gut, and act immediately" UnCollege.Org

"A child educated only at school is an uneducated child" George Santayana

"Children must be taught how to think, not what to think" The Idealist

"Re-examine all you have been told. Dismiss what insults your soul" Walt Whitman

A Theory of Everything

"What does school really teach you? Truth comes from authority. Intelligence is the ability to remember and repeat. Accurate memory and repetition are rewarded. Non-compliance is punished. Conform: Intellectually and socially" Unknown

"Education is not the learning of the facts, but the training of the mind to think" Albert Einstein

"When it comes down to it, the only knowledge that really matters is how to purify water, how to grow your own food, how to cook, how to build, and how to love. And funnily enough, we're not taught any of it in school... It's almost as if they want your head filled with crap that will never benefit your life, so you are always dependent on the government and corporations" Unknown

"The school system is designed to teach obedience and conformity and prevent the child's natural capacities from developing" Noam Chomsky

"Formal education will make you a living; self-education will make you a fortune" Jim Rohn